Variant Modeling with SysML

- MBSE4U Booklet Series -

Tim Weilkiens

Variant Modeling with SysML

- MBSE4U Booklet Series -

Tim Weilkiens

ISBN 978-3-9817875-7-3

*/***MBSE4U**

MBSE4U - Tim Weilkiens is a publishing organization for books about MBSE. They are intended to be regularly updated to align the content with the highly dynamic systems engineering domain.

Contents

About MBSE4U

MBSE4U - Tim Weilkiens is my publishing organization for MBSE books. The main focus is eBooks that are regularly updated to follow the dynamic changes in the MBSE community and the markets. Printed books are also available.

MBSE4U has published

- Tim Weilkiens. SYSMOD - The Systems Modeling Toolbox - Pragmatic MBSE with SysML. http://leanpub.com/sysmod. 2015. (ISBN 978-3-9817875-1-1 (PDF), 978-3-9817875-2-8 (ePub), 978-3-9817875-3-5 (MOBI), 978-3981787504 (Print))
- Tim Weilkiens. Variant Modeling with SysML. http://leanpub.com/vamos. 2016.
- Tim Weilkiens, The New Engineering Game. http://leanpub.com/new-engineering-game. Planned 3Q 2016.
- Tim Weilkiens, MBSE Craftsmanship. http://leanpub.com/mbse-craftsmanship. Planned 1Q 2017.

⚒MBSE4U

About me

I am a managing director of the German consulting company oose, a consultant and trainer, and active member of the OMG and INCOSE community. I have written sections of the initial SysML specification and I am still active in the ongoing work on SysML. I am involved in many MBSE activities and you can meet me on several conferences about MBSE and related topics.

As a consultant I have advised a lot of companies in different domains. The insights into their challenges are one source of my experience that I share in my books and presentations.

I have written many books about modeling including *Systems Engineering with SysML* (Morgan Kaufmann, 2008) and *Model-Based System Architecture* (Wiley, 2015). I am the editor of the pragmatic and independent MBSE methodology SYSMOD – the Systems Modeling Toolbox (We15).

You can contact me at tim@mbse4u.com and read my blog about MBSE at www.model-based-systems-engineering.com.

History and Outlook

This chapter gives a brief overview about the version history of the variant modeling approach and looks forward to future plans.

Version History

- 1.0 Initial comprehensive publication of VAMOS.

Outlook

- More analysis of other variant modeling approaches
- Provide more support of the profile for common SysML modeling tools.
- Explicit description of the alignment with other variant modeling approaches.
- Provide a guideline and template for report generation

Preface

From the early beginning of SysML I consult organizations from different domains to apply model-based systems engineering (MBSE) methodologies. Very often one of the first questions after the basics were understood is how to model variants with SysML. There is a high and increasing demand for variant modeling. The SysML does not provide first class model elements for variant concepts. However general SysML model elements like the *Generalization* relationship could be used to build a system model with variants.

I have developed a method how to model variants with SysML and have presented it in different publications, conference talks, or blogs. The original source of the concepts provided in this book is the MBSE methodology SYSMOD (We15). I first described the variant modeling method in the book *Systems Engineering with SysML/UML* (We08, We14). This book is a more comprehensive and recent description and is now the official source of the variant modeling method. Because it is much easier to talk about something that has a name I call the VAriant MOdeling method for SysML *VAMOS*.

The concept of VAMOS is not a complete new creation of mine. It is more a new configuration and adaption of existing concepts for the usage with SysML. Chapter 4 covers some other variant modeling concepts.

A disadvantage of classical books is the low frequency of updates. That was my main motivation to publish this book about variant modeling with SysML with my own publishing organization MBSE4U. I continuously update the method based on feedbacks and experiences of industrial projects and changes in the context like other new variant methods or changes of SysML.

A disadvantage of self-published books is the missing quality gate of a traditional publisher. There is no copy-editor who, for example, proves the correct usage of the English language - in particular whether the author is not a native speaker like me - or whether the line of arguments makes sense to the readers.

I appreciate any feedback on the book. Be it on the content or on my English skills. You can reach me by email: tim@mbse4u.com.

Thanks to Keith Smith who gave me a lot of feedback about SYSMOD, FAS and VAMOS.

I thank my colleagues at my company oose, in particular Axel Scheithauer and Stephan Roth for long profound discussions about MBSE.

I thank NoMagic for their support. I have created the SysML diagrams in this book with their modeling tool Cameo Systems Modeler.

Finally I would like to thank you for buying this book. The money is well spent. Now you have a description how to model variants with SysML. And I have some money to finance the infrastructure to provide MBSE goodies to the MBSE community like my blog www.model-based-systems-engineering.com.

If you need MBSE training or consulting services feel free to contact me. My company - the consultancy oose - provides MBSE trainings and coaching, for example to introduce MBSE in your organization.

Tim Weilkiens, April 2016.

1. Introduction

All along products exist in different variants. In recent years organizations face more and more the challenge to provide a huge set of product variants. The industry moves from the phase of mass production to a new phase of mass customization.

There are many reasons to manage variants in a model: A product line, a customized product or different designs for trade studies. Typically a single variant of a system affects only a few parts of the system. It is a slight derivation from the initial system. However it is not possible to quantify the number or level of detail that is allowed to vary to be still a variant of a system and not a new system.

You cannot measure abstraction and you cannot give an objective metric for the deviation of variants. The benefit must be larger than the effort to manage a complex variant model.

A car as well as an aircraft could be a variant of a transportation system. However in most cases it makes no sense in practice to handle a car and an aircraft as variants of a transportation system with all the appropriate relationships in a single system model. The common parts of a car and an aircraft are too abstract.

The description of variants is a sophisticated task. It is already challenging to create a good description of a single system. Every variation adds another dimension to a multi-dimensional system model. For example, the engine could be a variation of a car system with three possible variants: diesel, electric, or hybrid engine. Next variation could be the chassis with the variants small, deluxe, cabriolet. Now you can combine the variants, for example a car with diesel engine and a small chassis or a car with a hybrid engine and a deluxe chassis, and so on. Any additional variation increases the dimension and the number of potential combinations.

The OMG Systems Modeling Language (SysML)[1] (Sy15) is a general-purpose modeling language for systems engineering applications. It is mainly used for requirements and system architecture specifications. Both are strongly affected by variants. Therefore, it is highly valuable to use SysML for the specification of variants.

The book implies knowledge about SysML. If you need more information about SysML I recommend my book *Systems Engineering with SysML/UML* published by Morgan Kaufman (We08, We14) or for experienced readers the SysML specification itself (Sy15).

In the following chapter I give some definitions of variant terms. The next chapter describes the concept for VAriant MOdeling with SysML (VAMOS).

To think out of the box chapter 4 covers other variant modeling concepts. Finally chapter 5 presents the definitions of the VAMOS stereotypes.

The appendix provides additional examples of VAMOS: a forest fire detection and a virtual museum tour system.

---

[1]OMG SysML is a trademark of the Object Management Group.

2. Variant Modeling Concepts

In this chapter I give the definitions of variability terms how I use them in the context of variant modeling to get a solid fundament. The terms and concepts are conform with common variant concepts presented in publications about variant modeling, for example the Orthogonal Variability Model (OVM) (Po05) (see chapter 4).

Figure 2.1[1] shows the conceptual model of the VAMOS terms. The stereotype *«domainBlock»* represents conceptual elements. It is part of the SYSMOD profile (Wei15). A conceptual model[2] is a helpful, comprehensive, concise and clear description of domain terms.

On the top level I differentiate between the core, the variants, and the configurations.

The core contains all elements that are used in all system configurations. A **core element** is an element of the core and independent of any variant element.

A **variation point** marks a core element of the system as a docking point for a variant element.

A **variant element** only occurs in some configurations and is part of exactly one variant. The variants and the core are orthogonal concepts. The core is independent of the variants. The variants could be stored in a separate physical model.

---

[1]The colored boxes in the diagram are no SysML model elements. They are geometric figures to emphasize the separate variant concepts. Most modeling tools provide those graphical adornments of SysML diagrams.

[2]In SYSMOD the conceptual model is named domain knowledge model.

A **variant** is a complete set of variant elements that varies the system according to a variation. A variant is also known as a feature of the system.

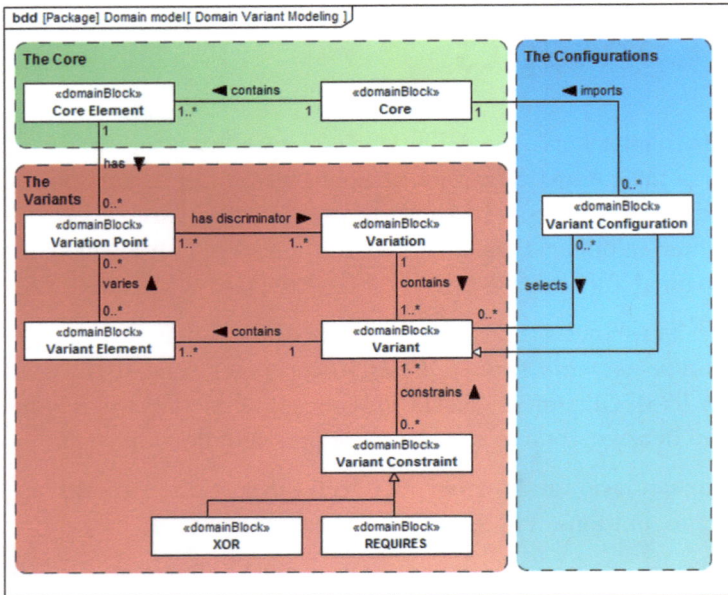

Figure 2.1: **Variant Modeling Domain**

A **variation** is the discriminator for variants. For example, diesel, electric, and hybrid engines are variants of the variation engine kind.

A variant could again include variations (figure 2.2). The structure of these variations is the same as for the top level variations. This recursive structure makes the variant modeling concept scalable for any size of a system.

The core is a concept relative to the variations on the same level. A variant that again includes variations contains also core elements relative to these variations. But they are variant elements relative to the variations of the upper level.

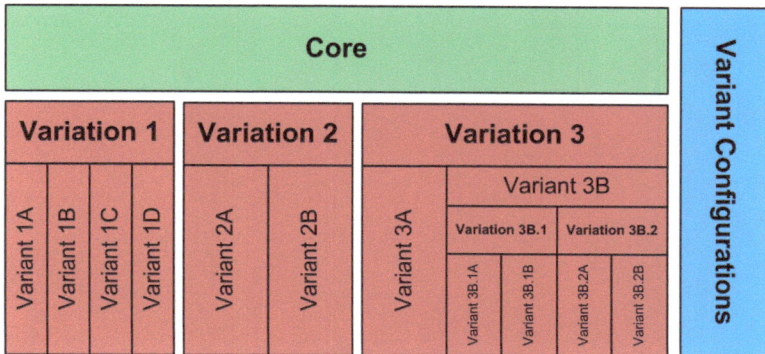

Figure 2.2: Recursive Structure of a Variant Model

A **variant configuration** is a valid set of variants and the core, for example, a car with a hybrid engine and a deluxe chassis. A variant configuration is also a special variant and part of a variation. In our example the variant configuration *Hybrid engine with Deluxe chassis* is part of the variation *eco editions*.

A **variant constraint** specifies rules for a valid set of variants. Two common variant constraints are predefined. XOR is used to exclude a variant if a specific variant is selected. REQUIRES specifies that other variants are required if a specific variant is selected.

The SysML profile for VAMOS presented in chapter 5 implements the concepts.

3. Variant Modeling with SysML (VAMOS)

SysML does not provide explicit built-in language constructs to model variants. Nevertheless SysML is useful to create a model with variability aspects. The VAMOS method presented in this chapter is one option how to model variants with SysML. It uses the profile mechanism of SysML to extend the language with a concept for variant modeling. The chapter VAMOS Stereotypes describes the stereotypes for the variant modeling concepts. The stereotypes extend SysML to model the concepts of variants, variations, variation points, variant elements, variant configurations and variant constraints presented in chapter 2. The VAMOS Stereotypes are part of the SYSMOD profile (We15), although they could be used independently of SYSMOD.

Typically modeling in general has three separate concerns: the method, the language, and the tool (figure 3.1).

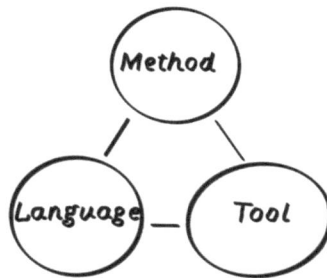

Figure 3.1: The Three Concerns of Modeling

Mapped to the context of VAMOS, VAMOS itself is a method and the VAMOS stereotypes are part of the language. The modeling tool for

VAMOS could be any modeling tool that supports standard SysML.

A great benefit of the VAMOS approach is that there is no need to stress a 3rd party tool for the variant modeling. Certainly a tool with specialized handling of variants has benefits. However the price is another tool including license costs, training, learning yet another modeling language, a tool chain between the SysML tool and the variant tool, a cumbersome engineering environment, and more. It finally depends on your requirements on the variant modeling which approach has the highest value for you. If you have very ambitious requirements, it could makes sense to use a highly specialized tool for variant modeling. Otherwise the value of using VAMOS with a standard SysML modeling tool is higher (figure 3.2).

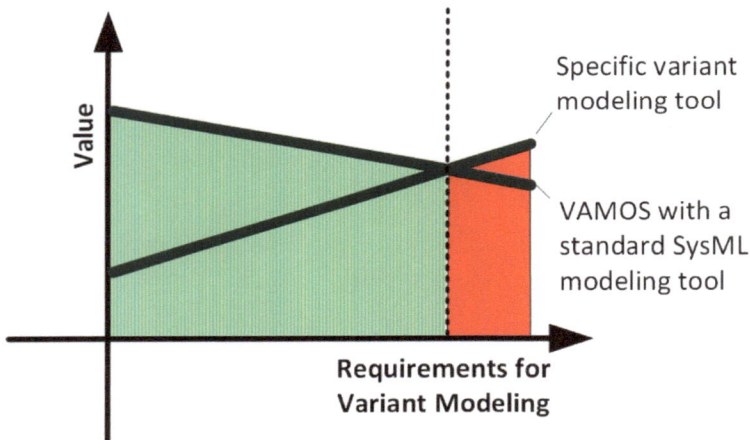

Figure 3.2: Value of VAMOS

Note that the VAMOS line in figure 3.2 stands for VAMOS with a standard SysML modeling tool. If you customize the modeling tool and add specific variant modeling functionality you can jump on the higher line at the right side of the figure. In that case it is not a standard SysML modeling tool anymore, but also a specific variant modeling tool.

3.1 Sample Project

In this chapter I use a special example to demonstrate the concept. It is a cookbook for fruit salads. That way you can focus on the variant modeling method and the usage of SysML and you are not distracted by a technical example. More seriously you find examples of technical systems in the appendix of this book (Virtual Museum Tour, Forest Fire Detection System).

The *Fruit Salad* model describes some fruit salad recipes taken from the cookbook website *allrecipes.com* (Al16). A single recipe is a variant configuration based on core elements and a set of selected variants, for example a special kind of an apple, orange, or dressing.

3.2 Core, Variations, and Variant Configurations

The basic concepts of VAMOS are the core, the variations, and the variant configurations (see also chapter 2). This section gives a general overview. Detailed descriptions of the three concepts are given in following separate sections.

Figure 3.3[1] shows the top level package structure of the *Fruit Salad* model. The model represents the whole configuration space as well as some concrete variant configurations.

---

[1]The colored boxes in the diagram are no SysML model elements. They are geometric figures to emphasize the separate variant concepts. Most modeling tools provide those graphical adornments of SysML diagrams.

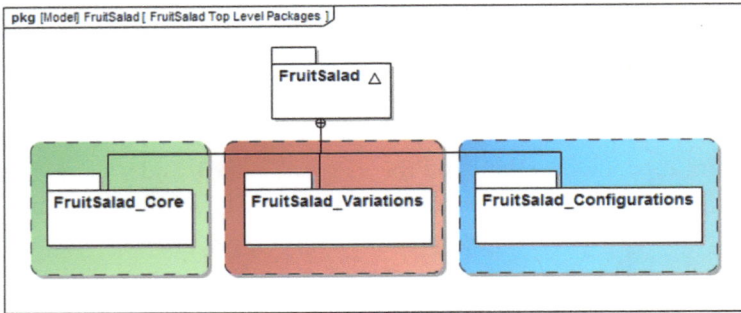

Figure 3.3: Top Level Packages of the *Fruit Salad* Variant Model

On the first level of the model we have three packages that represent the basic concepts:

- The Core package (*FruitSalad_Core*) contains all core elements. The structure of the sub-packages conforms to the system model structure presented in We15. However you can use any package structure here. The structure is independent of the variant modeling approach. Figure 3.4 depicts the top level package structure of the *Core* package. Section 3.3 gives a more detailed description of the core.

- The Variations package (*FruitSalad_Variations*) contains all variations with their variants. The feature tree in figure 3.15 depicts four variations: *Pepper, Apple, Orange*, and *Dressing*. A variation is a package with stereotype *«variation»*. Each variation package contains the variants according to the variation discriminator. Section 3.5 and section 3.6 give a more detailed description of the variations.

- The Configurations package (*FruitSalad_Configurations*) contains concrete variant configurations, i.e. valid sets of core and variant elements combined to a system or system assembly respectively a fruit salad or fruit salad ingredient. Since a variant configuration is also a special variant (see figure 2.1) we have variation packages on the first level and the variant configurations as variants on the next level. The variant configuration package includes one variation *Recipes* with two variant configurations: the *Recipes_JuicyFruitSalad* and *Recipes_AppleColeSlaw* (Figure 3.5). Section 3.7 gives a more detailed description of variant configurations.

This package structure clearly separates the three orthogonal variant concepts (figure 3.4).

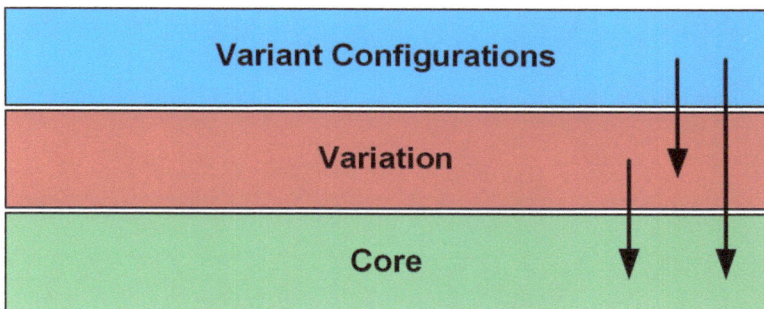

Figure 3.4: Orthogonal aspects

The variant configurations depend on the variant and the core assets. The variation assets depend only on the core assets. And the core assets are independent of the variation assets and variant configuration aspects despite the information about variation points.

3.3 Core

This section gives a brief overview of the core. It is a "normal" system model and independent of the variant aspects except the assignment of variation points (see section 3.4). You can use any methodology to create the core model. I have used my methodology SYSMOD described in We15 to create the core.

Figure 3.5[2] depicts the top level package structure of the core. The notion of the package structure is described in We15. It is a best practice for SysML models, but independent and not mandatory for the application of VAMOS.

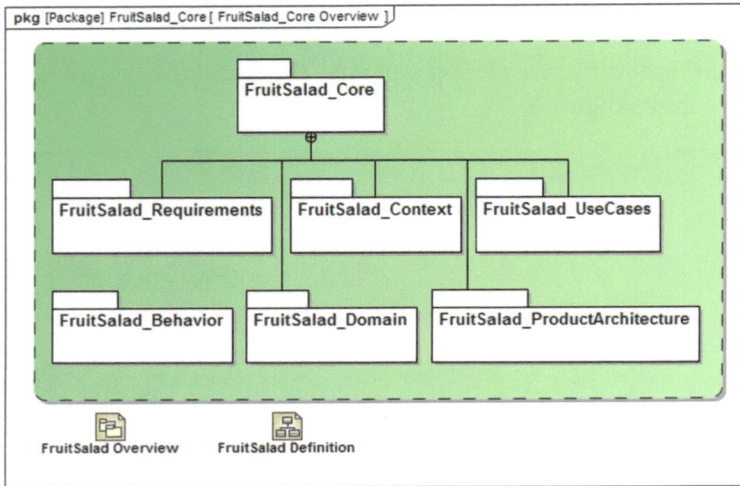

Figure 3.5: **Top Level Packages of the *Core* package**

Figure 3.6 depicts the root system element *Fruit Salad* with the system idea. The text of the system idea is stored in the property *systemIdea* of the SYSMOD stereotype «*system*».

Figure 3.7 shows the objectives of the *Fruit Salad*. The *Tasty* objective directly addresses the salads while the *Learning* objective is more a meta-objective that addresses the VAMOS approach.

---

[2]The *FruitSalad Overview* and *FruitSalad Definition* elements in the diagram are hyperlinks to a SysML diagram of same name. Diagram hyperlinks are not part of SysML and not mandatory for VAMOS (or SYSMOD). Most SysML modeling tools provide that feature and it is convenient to provide navigation paths through the model.

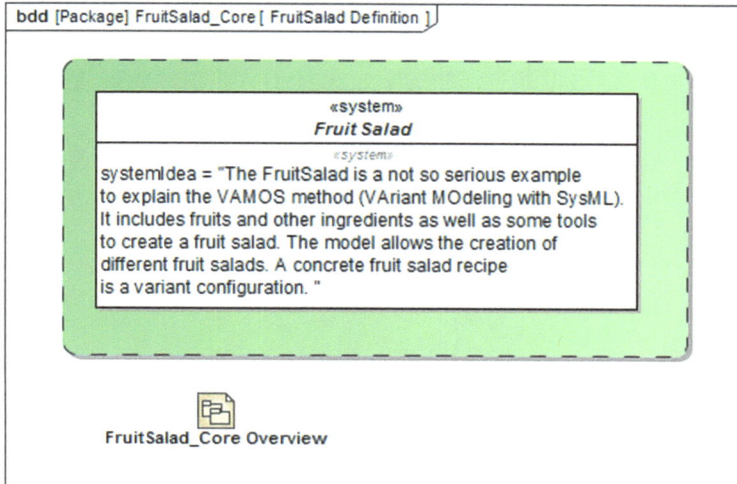

Figure 3.6: Fruit Salad System Idea

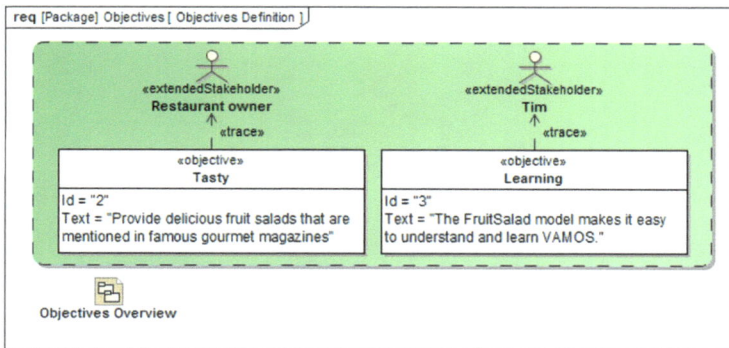

Figure 3.7: Fruit Salad Objective

Figure 3.8 depicts the system context of the fruit salad. The model does not cover how the ingredients of the fruit salad are obtained. Therefore there is no appropriate supplier actor and item flow in the context.

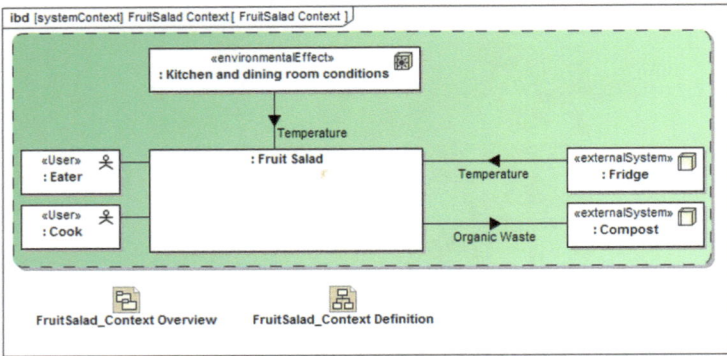

Figure 3.8: Fruit Salad System Context

The user *Eater* eats the fruit salads and the user *Cook* creates them. Figure 3.9 depicts the main use cases of the users.

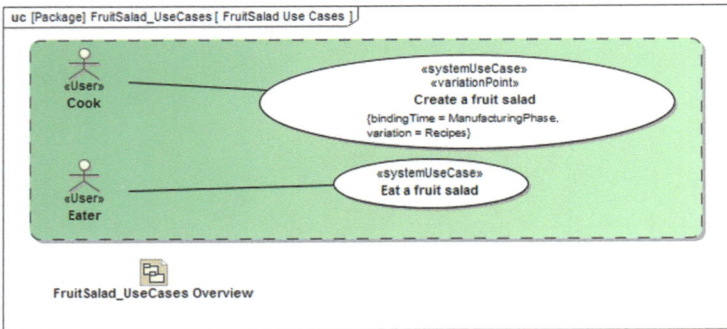

Figure 3.9: Fruit Salad Use Cases

Figure 3.10 lists some non-functional requirements. The functional requirements are covered by the use cases (figure 3.9).

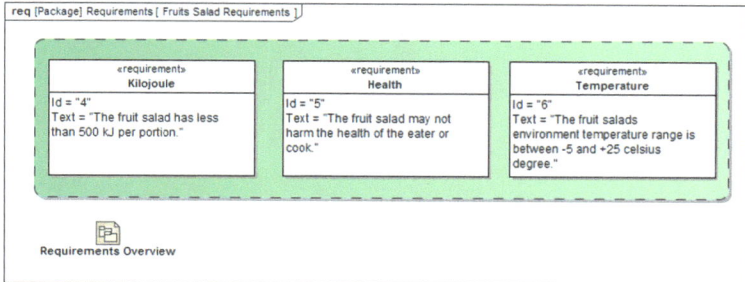

Figure 3.10: Some Fruit Salad Requirements

Figure 3.11 depicts elements of the fruit salad domain knowledge.

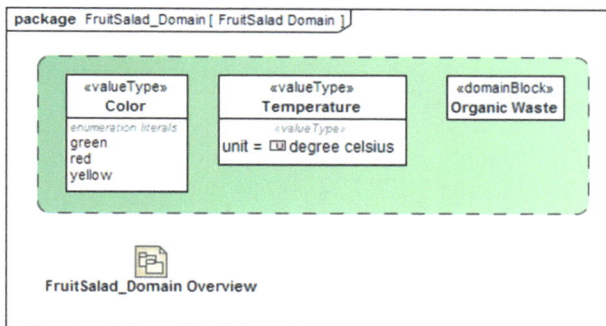

Figure 3.11: Fruit Salad Domain

The fruit salad uses a simple cooking model library that could also be used by other models (figure 3.12).

Many blocks of the fruit salad product architecture inherit from the abstract block *Ingredient* from the model library (figure 3.13). The model library also provides some concrete ingredients, a bowl, and the data type *Size*. These are just some examples. Of course there could be much more in a model library for cooking recipes.

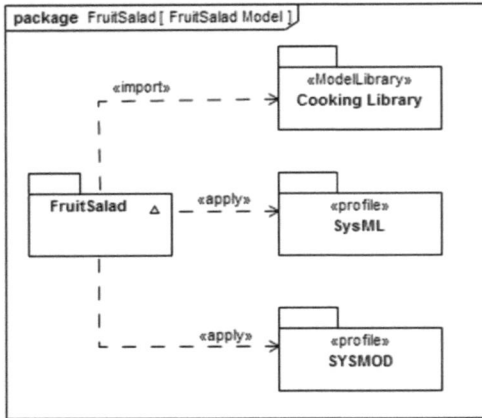

Figure 3.12: Fruit Salad Model

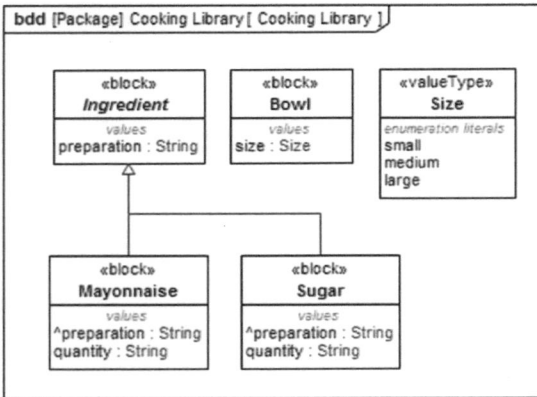

Figure 3.13: Cooking Model Library

Figure 3.14 depicts elements of the product architecture. The fruit salad system is separated in the subsystems *Fruit Subsystem, Vegetable Subsystem, Other Ingredients* and *Tools*.

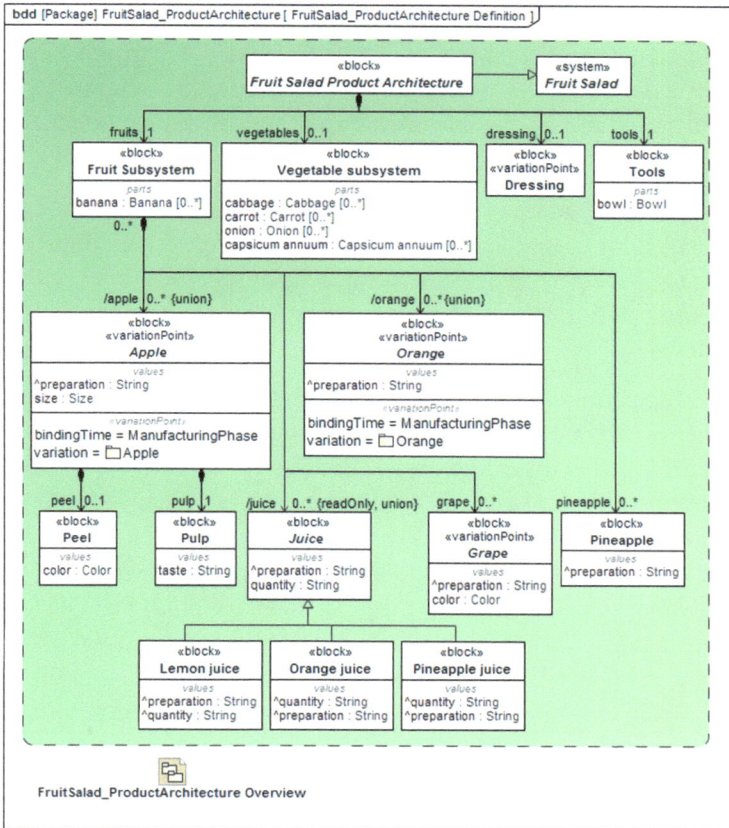

Figure 3.14: Fruit Salad Product Architecture Definition

The core of the fruit salad does not specify a specific system. It is a toolbox that allows many different kinds of system respectively fruit salads. Therefore there is more specification work to do in the variant configurations to define a concrete single fruit salad (see section 3.7).

Interestingly some of the blocks of the product architecture could also be domain blocks, for example the *Apple* or *Orange*. Typically domain blocks are not part of the architectures. Here I have decided to model *Apple* & Co. as elements of the product architecture and not as parts of the domain knowledge. Whereas the *Organic Waste* block is a domain block and is not part of the product architecture. It is only used in the system context and the use case model of the system.

Not any specialization of core classifiers must lead to a modeling of variants. For example, figure 3.14 depicts different kind of juices. All of them are part of the core and the general block *Juice* is not a variation point. It is up the modeler to decide to make variants explicit.

Some properties in figure 3.14 are special and need some more explanations. For example the property *apple* at the left side of the figure at the end of the association from the *Fruit Subsystem* block to the block *Apple*. The complete syntax of the property is */apple:Apple[0..*] {union}*. The well known elements are the property name (*apple*), the type (*Apple*) and the multiplicity (*[0..*]*). Special are the "/" sign and the property string *{union}*. The "/" sign indicates that the property *apple* is a so-called derived property. The property value is not directly stored in the appropriate slot, but is derived from other values in the model. The property string *{union}* specifies that the property is a union of sets. Those sets are properties marked with the property string *{subsets apple}*. You find them in specialized blocks, for example in figure 3.31.

The internal block structure plays not an important role for the fruit salads, although it is extremely important for most technical systems. You find examples of internal block structures of technical systems in the context of variant modeling in appendix A and appendix B. Nevertheless figure 3.15 depicts an internal block diagram of the core fruit salad system.

Talking about architectures often considers only the structural aspect. It is often overseen that the behavior aspect is also part of the architecture. In the fruit salad example many blocks of the product architecture owns a state machine behavior.

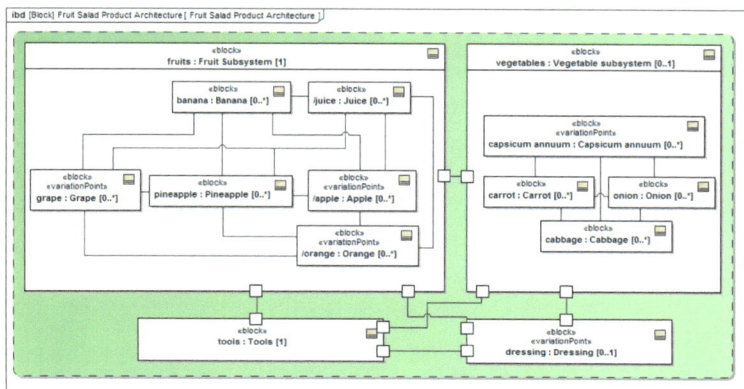

Figure 3.15: Internal structure of the Fruit Salad

Figure 3.16 depicts the state machine of the apple. For example, these states are used in the activities (figure 3.28 and figure 3.29).

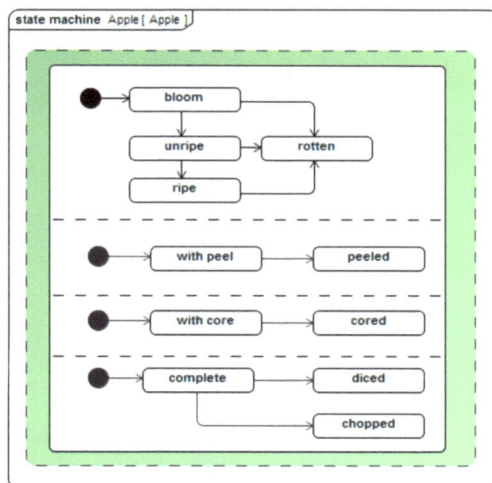

Figure 3.16: State machine of the Apple

3.4 Variation point

Variation points are elements in the core that are refined by variant elements. The *Apple* in the core is such a variation point. Figure 3.17 depicts the definition of the apple in the core. Variants specify different varieties of apples. The variation, i.e. the discriminator of the variants, is also named *Apple*. The binding time of an apple variant is the manufacturing phase, i.e. cooking time.

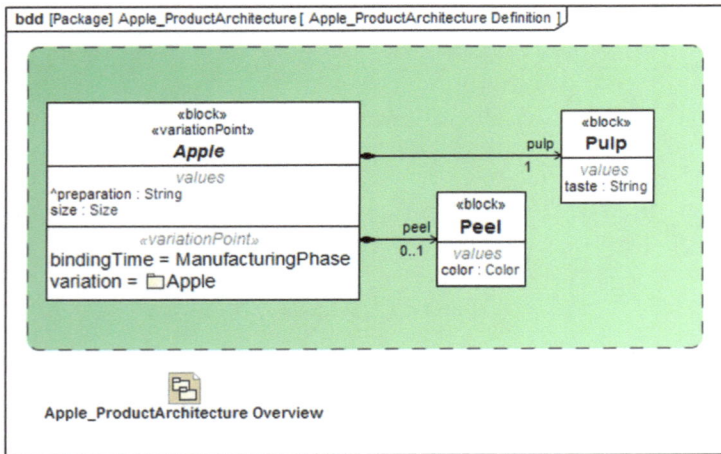

Figure 3.17: **Apple variation point in the Fruit Salad model**

The part types *Peel* and *Pulp* could also be variation points of the variation *Apple*. The specific apple variants have specific peels and pulps. However since the owner *Apple* of the peel and pulp is already a variation point, they are not explicitly marked also as variation points to minimize the modeling effort, although they are specialized by variant elements. Finally the decision is up to you how granular you model variation points.

Best Practice

Do not model the children of a variation point also as variation points, if they belong to the same variation. If a node is a variation point, all child nodes are implicitly covered by the variation point.

Section 3.6 describes the variants and how the apple variant elements are connected to the core elements.

Figure 3.18 depicts a table with all variation points in the core. Most modeling tools provide table views like this. The table gives a good overview and shows always all available information. It is created automatically by a model query and not manually by a modeler like a standard SysML diagram that only depicts the elements that the modeler has placed on it.

#	Owner	Name	Variation	Binding Time	Specializations
1	FruitSalad_ProductArchitecture	Orange	Orange	ManufacturingPhase	Common Orange / Blood Orange
2	FruitSalad_ProductArchitecture	Grape	Grape	ManufacturingPhase	Green Grape / Red Grape
3	FruitSalad_ProductArchitecture	Dressing	Dressing	ManufacturingPhase	Sweet Mayo Dressing
4	FruitSalad_UseCases	Create a fruit salad	Recipes	ManufacturingPhase	Create a Apple Cole Slaw / Create a Juicy Fruit Salad
5	Create a fruit salad	Create a fruit salad	Recipes	ManufacturingPhase	Create a Apple Cole Slaw / Create a Juicy Fruit Salad
6	FruitSalad_ProductArchitecture	Capsicum annuum	Capsicum Annuum	ManufacturingPhase	Bell Pepper / Chilli Pepper
7	Apple_ProductArchitecture	Apple	Apple	ManufacturingPhase	Granny Smith / Cortland / Aurora Golden Gala

Figure 3.18: Overview Variation points in the Fruit Salad model

Best Practice

Do not focus too much on diagrams. Graphical views are very powerful, but that does not mean that you should use them everywhere. If you want to show a list of things, use a table view instead of a diagram. If you want to show the relationships between two element types, use a matrix view.

Table and matrix views are typically based on a model query and show the complete set of information regarding the query. A standard SysML diagram shows only the information that the modeler has placed on it.

The column *Specializations* in figure 3.18 lists all elements that have a generalization relationship to the specific variation point. The entry makes only sense for variation points applied to a classifier; otherwise the variation point could not be the target of a generalization relationship.

3.5 Variations and Feature Trees

Feature trees are a common representation of variant options. Other concepts that use feature trees are for example the *Orthogonal Variability Model* (OVM) or the *Feature Oriented Domain Analysis* (FODA) (see chapter 4).

The package structure of the VAMOS method enables a feature tree representation in a SysML package diagram. Figure 3.19 depicts parts of the feature tree of the *Fruit Salad*. The root element of the tree is the top level variations package. The next level lists the variations depicted by packages with stereotype *«variation»*. And the leaves of the feature tree are the variant packages. A variant also represents a feature of the system.

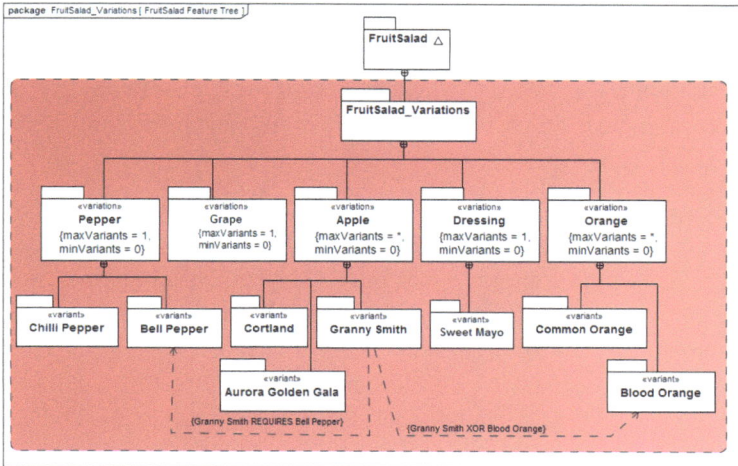

Figure 3.19: Feature Tree of the Fruit Salad

Each variation package has two additional properties (*minVariants* and *maxVariants*) that constrain the number of variants of the variation that could be selected for a single valid variant configuration.

In Figure 3.19 the *minVariant* and *maxVariant* specifications are depicted in the package symbol of the variations. For example, the *Pepper* variation allows only one variant per variant configuration (*maxVariants=1*). For example, if *maxVariants* would be 2 it is allowed to use two different kinds of pepper in a variant configuration. The *Pepper* variation is optional, i.e. a valid variant configuration could have no pepper. That is specified by *minVariants=0*. The *Orange* and *Apple* variations allow an unlimited number of selected variants depicted by the '*' symbol (=unlimited value of the UnlimitedNatural data type). It could also be zero.

The feature tree also depicts rules between variants. XOR and REQUIRES are special variant constraints to model rules between variants of the same or different variations. They are stereotypes to be applied at the SysML *Constraint* model element (see section 5.2 and section 5.7).

The XOR constraint specifies a mutual exclusive relationship between two variants. Figure 3.19 depicts a XOR between the variants *Granny Smith* and *Blood Orange*. Although some people might like that combination in a fruit salad, this cookbook model prohibits fruit salads that contain both.

The REQUIRES constraint specifies that the selection of a variant requires the additional selection of another variant for a valid variant configuration. In figure 3.16 the variant *Granny Smith* requires the variant *Bell Pepper*. The REQUIRES constraint is a unidirectional constraint. *A requires B* does not imply that B also requires A. The direction of the constrained elements is defined by the order of the operands. The property *constrainedElement* of the SysML *Constraint* is an ordered list. In figure 3.19 the order is depicted by an arrow at the dashed constraint anchor line.

The XOR and REQUIRES constraints are predefined constraints of VAMOS. They are the most common constraints for feature trees. You can define additional constraints by using the SysML model element *Constraint*.

The VAMOS stereotypes define OCL rules for the variant constraints (see chapter 5). If your modeling tool provides OCL, it can validate the constraints, i.e. automatically validates if a variant configuration imports a valid set of variants. Otherwise it is only information.

The variant constraints could also be shown in a matrix. The matrix in figure 3.20 lists all variant constraints in the rows and the constrained elements in the columns. The matrix reveals that it is useful to give the constraints a good name, here *Granny Smith REQUIRES Bell Pepper* and *Granny Smith XOR Blood Orange*.

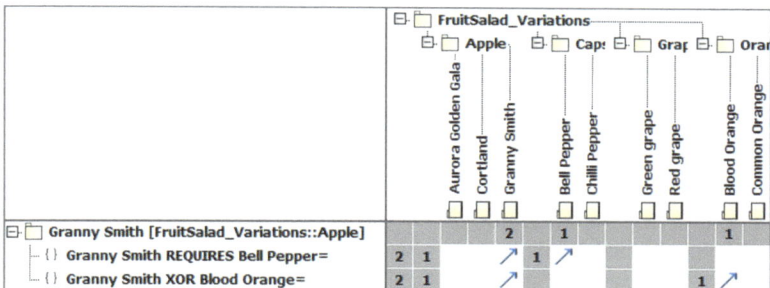

Figure 3.20: **Variant Constraints Matrix**

3.6 Variants

A variant is a package with stereotype *«variant»*. The variant package is the root for all elements of a single variant. The elements are organized like the recursive package structure for system models (We15). A variant could be handled like a system or subsystem with its own context, requirements, architectures as well as again configurations, variations and variants.

Figure 3.21 depicts the package structure of the apple variants. Each variant contains only a package for product architecture elements.

Figure 3.22 depicts elements of the product architecture of the variant *Aurora Golden Gala* and their relationships to core elements.

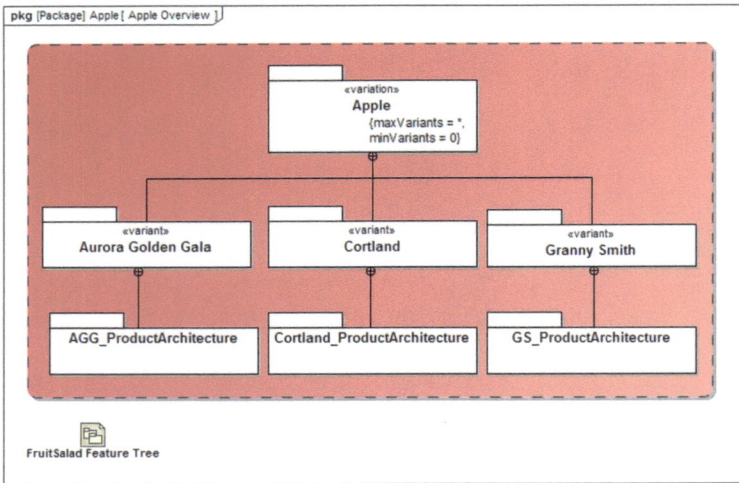

Figure 3.21: Package structure of *Apple* variation

The *Aurora Golden Gala* is a special kind of an *Apple*. It also has a special peel *AGG Peel* and a special pulp *AGG Pulp*. The *AGG Pulp* inherits the property *taste* from the core element *Pulp*. The *AGG Pulp* property *taste* with the initial value "sweet" must redefine the inherited property *taste*; otherwise the block *AGG Pulp* would have 2 properties with name *taste*: the inherited one and the newly defined one with the initial value. The same is true for the property *color* of the *AGG Peel* block and the properties *peel* and *pulp* of the *Aurora Golden Gala* itself. For example, the property *pulp* at the core element *Apple* is of type *Pulp* while the new property *pulp* of the variant element *Aurora Golden Gala* is of type *AGG Pulp*.

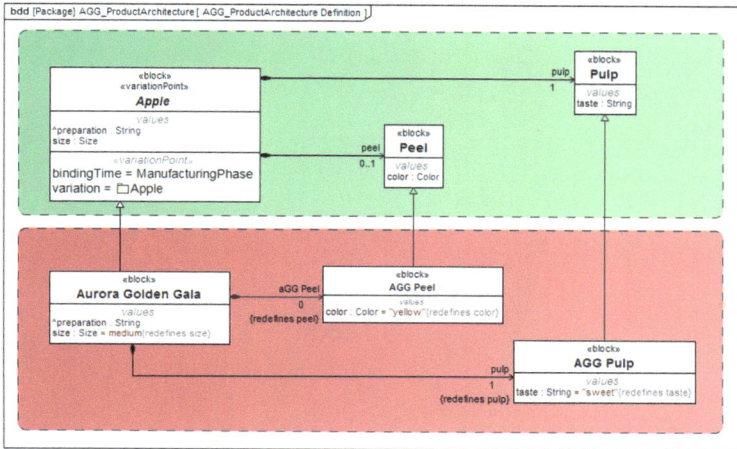

Figure 3.22: Relationship between variant and core elements

Stereotypes and their properties (tagged values) are not inherited. Therefore the block *Apple* is a variation point while the specialized block *Aurora Golden Gala* is not a variation point.

The Variant Interface Matrix depicted in figure 3.23 shows all relationships of variant elements to core elements. It is a helpful view on the model to manage the interfaces between the variant and the core.

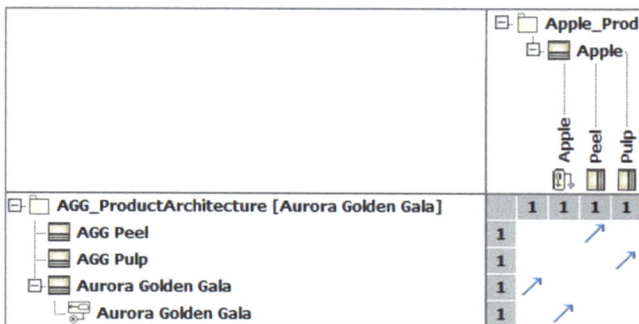

Figure 3.23: Variant Interface Matrix

Another variant is the *Sweet Mayo Dressing* (see feature tree in figure 3.19). Figure 3.24 depicts the product architecture definition of the dressing.

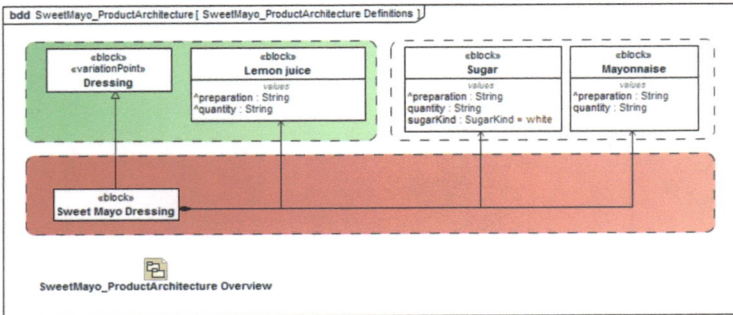

Figure 3.24: Product architecture of the Sweet Mayo Dressing

The block *Sweet Mayo Dressing* is a specialization of the variation point *Dressing*. The parts of the dressing are typed by blocks from the core (*Lemmon juice*) and by blocks from a model library (*Sugar* and *Mayonnaise*).

Figure 3.25 depicts the internal structure of the *Sweet Mayo Dressing*. The values of the properties *quantity* and *sugarKind* are so-called context-specific values. The values only exist in the context of the block *Sweet Mayo Dressing* and could be different in another context.

Figure 3.25: Internal structure of the Sweet Mayo Dressing

Context-specific values are a special feature of SysML. They are very useful. Unfortunately I do not know many modeling tools that provide the modeling of context-specific values and I have rarely seen them in real projects. An alternative for context-specific values is to use generalization and redefinition of the appropriate properties like the special peel and pulp blocks of the Aurora Golden Gala (see above). However that leads to many more model elements that must be created and maintained. Another option is to use property-specific types. They are also a feature of SysML that is rarely implemented by tools and rarely used in real projects.

3.7 Variant Configuration

Figure 3.26 depicts the configuration tree. It is a special version of the feature tree presented in section 3.5. This example has only two variant configurations, i.e. fruit salad recipes.

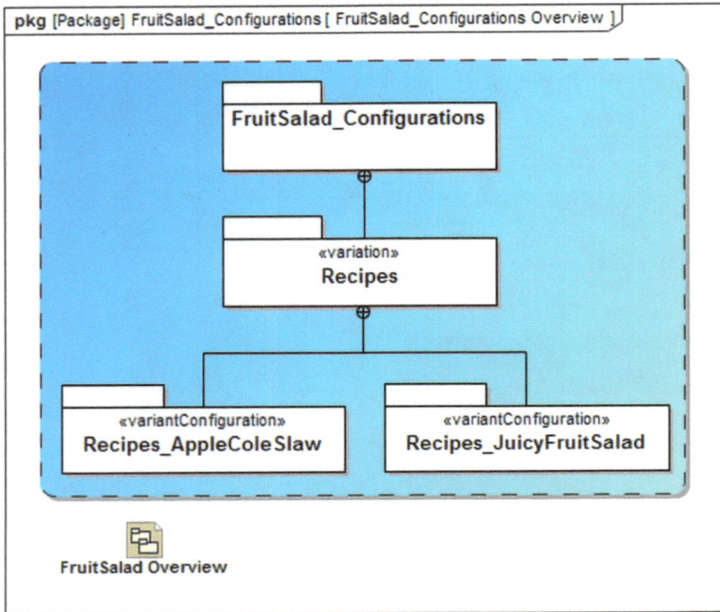

Figure 3.26: Variant Configurations

Figure 3.27 depicts the selection of the core and variants of the *Recipe_JuicyFruitSalad* variant configuration. In addition to it, the diagram depicts an image and the recipe of the juicy fruit salad. These are no SysML elements, but it is a nice feature to add such adornments to a diagram that most SysML modeling tools provide.

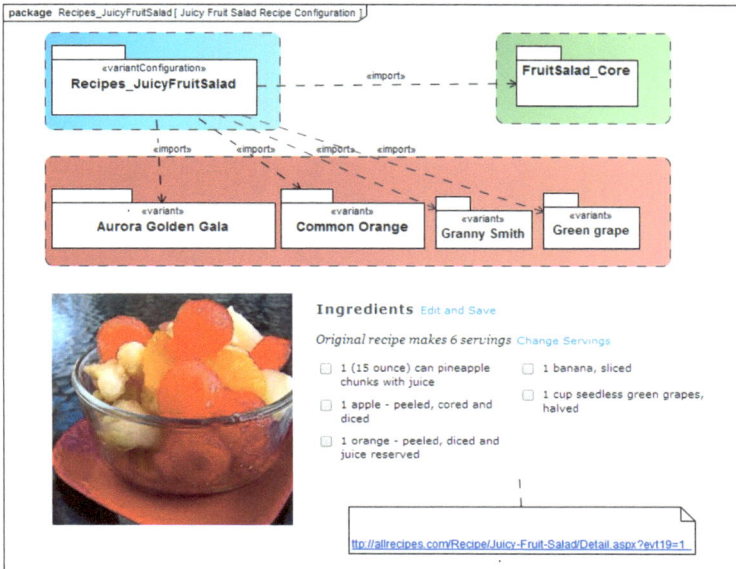

Figure 3.27: Recipe (variant configuration) Juicy Fruit Salad

The variant selection relationship between the variant configuration and the variants respectively the core is the standard SysML import relationship. In addition to the VAMOS semantic that the relationship selects the variant, it has its origin SysML semantic that the target namespace is imported into the source namespace. That means the selected variants are members of the variant configuration namespace.

Another more clear and comprehensive representation of the variant selection than figure 3.24 is a matrix as shown in the following figure 3.28. The matrix view depicts the result of a model query. The data is always complete. On the other side a SysML diagram shows only the elements that the modeler has put on the diagram. It is typically incomplete.

		Aurora Golden Gala	Cortland	Granny Smith	Bell Pepper	Chilli Pepper	Sweet Mayo	Green grape	Red grape	Blood Orange	Common Orange
Recipes [FruitSalad_Configurations]		1	1	2	1		1	1			1
Recipes_AppleColeSlaw	4	2	↗	↗	1	↗	1	↗			
Recipes_JuicyFruitSalad	4	2	↗		↗				1	↗	1	↗

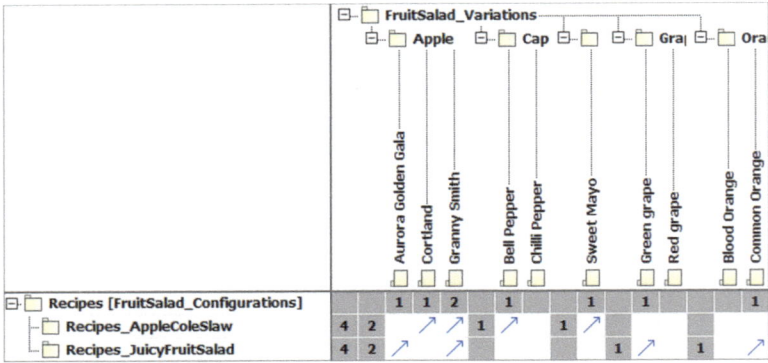

Figure 3.28: Variant Configuration Matrix

Figure 3.29 depicts the package structure of the variant configuration *Recipes_JuicyFruitSalad*. It has the same package structure as a variant package or the core package. Here the variant configuration *Recipes_JuicyFruitSalad* contains use cases, product architecture elements, and instance specifications.

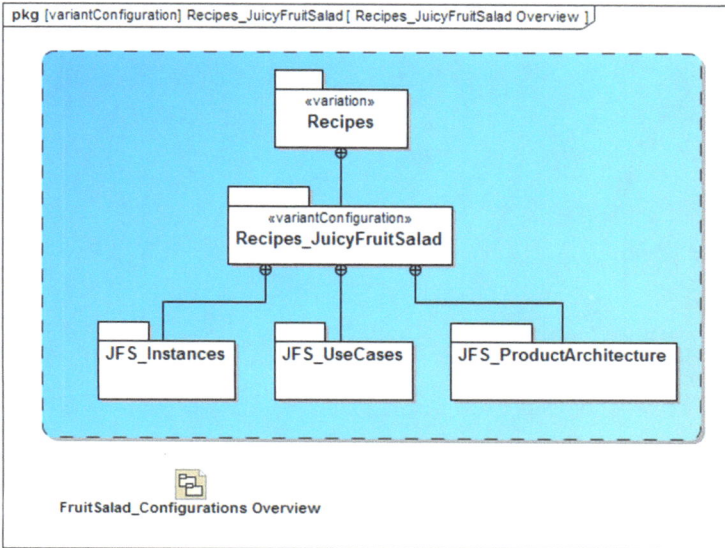

pkg [variantConfiguration] Recipes_JuicyFruitSalad [Recipes_JuicyFruitSalad Overview]

«variation»
Recipes

«variantConfiguration»
Recipes_JuicyFruitSalad

JFS_Instances   **JFS_UseCases**   **JFS_ProductArchitecture**

FruitSalad_Configurations Overview

Figure 3.29: Package Structure of variant configuration Juicy Fruit Salad

The use case package includes a single use case that represents the recipe. The use case is a specialization of the core use case *Create a fruit salad* (figure 3.30). The use case activity of the core use case is without further details like actions or control flows. The specialized use case activity describes all the details.

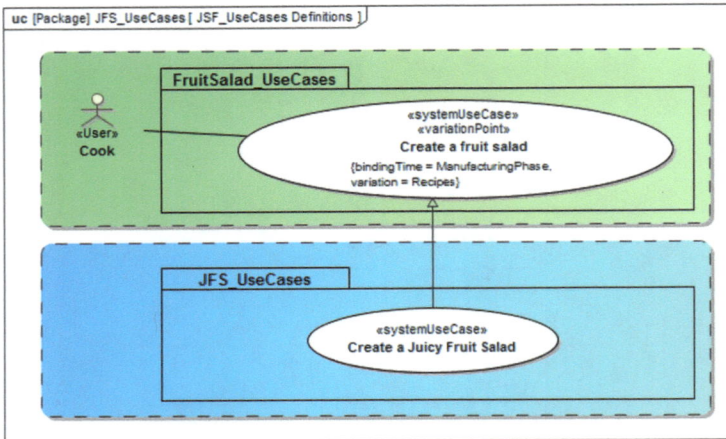

Figure 3.30: Use Case *Create a Juicy Fruit Salad*

The variant modeling of behavior touches a sore spot of the SysML. A key concept of variant modeling is the generalization relationship. It is allowed to model a generalization relationship between behavior elements like activities or state machines. However the responsible UML specification does not specify how to show inherited elements in a diagram, for example an activity diagram. In consequence no SysML modeling tool shows inherited elements. That makes it extremely difficult to model variants of behaviors. In addition to the notation issue the modeling of behavior generalization is an inherent sophisticated task.

I propose the following modeling guideline for behavior generalizations:

Best Practice

If the specialization differs only slightly from the general behavior, then model the details of the general behavior and describe the specialization only textually.

If the specialization differs strongly from the general behavior, then describe the general behavior textually and model the details of the specialized behavior.

Back to our fruit salad the general use case activity of the use case *Create a fruit salad* is described by text (figure 3.31).

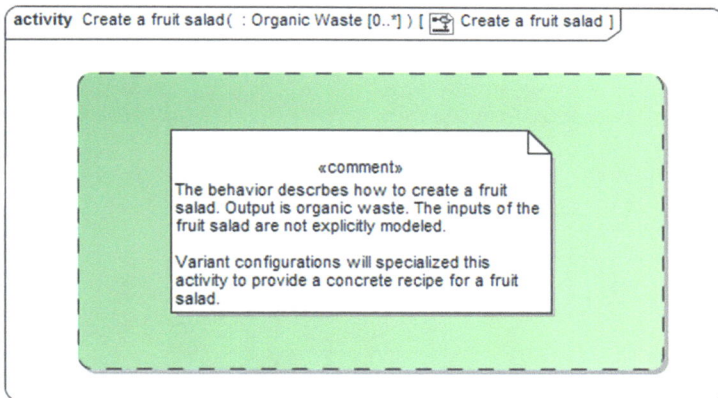

activity Create a fruit salad(: Organic Waste [0..*]) [Create a fruit salad]

«comment»
The behavior descrbes how to create a fruit salad. Output is organic waste. The inputs of the fruit salad are not explicitly modeled.

Variant configurations will specialized this activity to provide a concrete recipe for a fruit salad.

Figure 3.31: Use Case Activity *Create a fruit salad*

The specialized use case *Create a Juicy Fruit Salad* owns a use case activity of same name that specializes the general use case activity (figure 3.32).

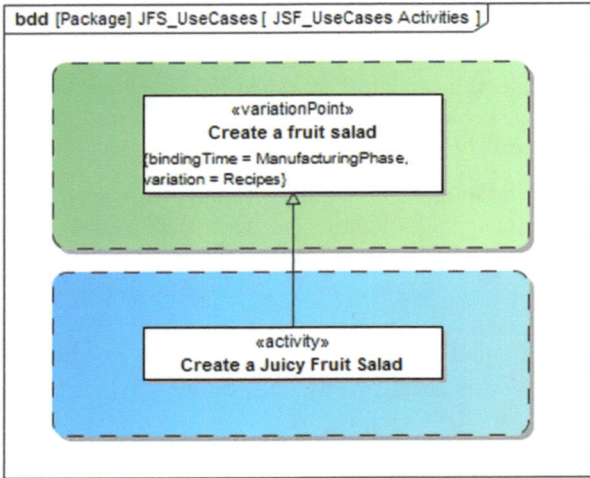

Figure 3.32: Specialization of Use Case Activity *Create a Juicy Fruit Salad*

The details of the activity *Create a Juicy Fruit Salad* are modeled and depicted in figure 3.33.

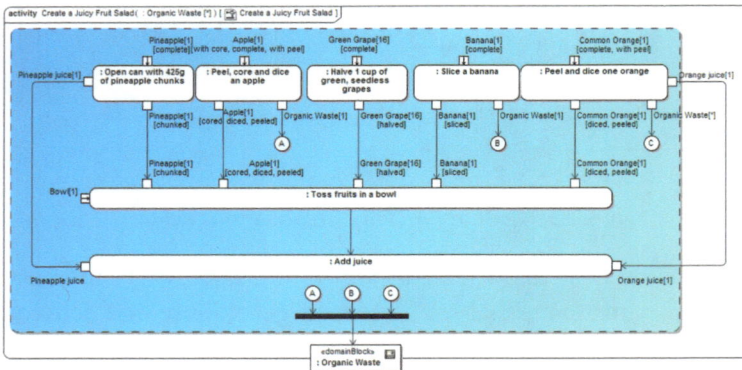

Figure 3.33: Recipe Juicy Fruit Salad Use Case Activity

The call behavior action to prepare the apple (*:Peel, core and dice an apple*) calls an activity defined in the core (figure 3.34). Figure 3.35

depicts the general apple preparation activities in the core. They are part of the core package *FruitSalad_Behaviors*.

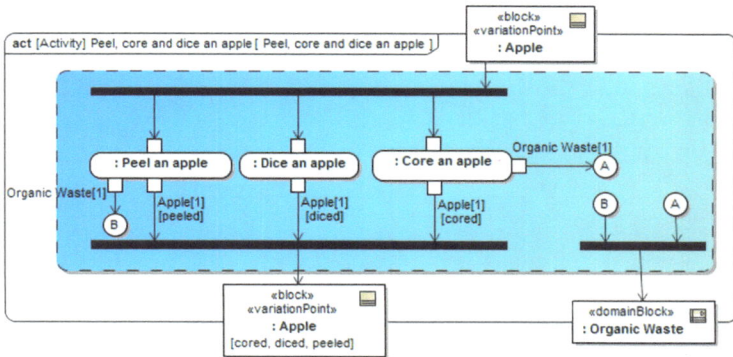

Figure 3.34: **Prepare apple for the Juicy Fruit Salad**

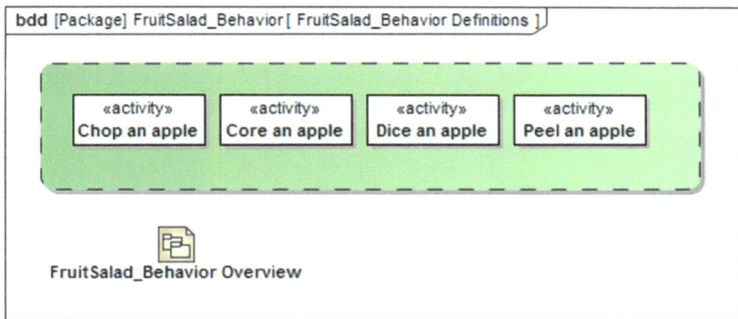

Figure 3.35: **Activities for the preparation of an apple**

Figure 3.36 depicts the product architecture of the juicy fruit salad. The specialized blocks *Juicy Fruit Salad* and *JFS Fruit Subsystem* redefine properties of the appropriate more general blocks of the core to remove properties or to specify the exact amount, for example zero vegetables or 16 grapes. In addition the redefined properties set specific fruit kinds like the green grape instead of grape or the *Aurora Golden Gala* instead of the apple.

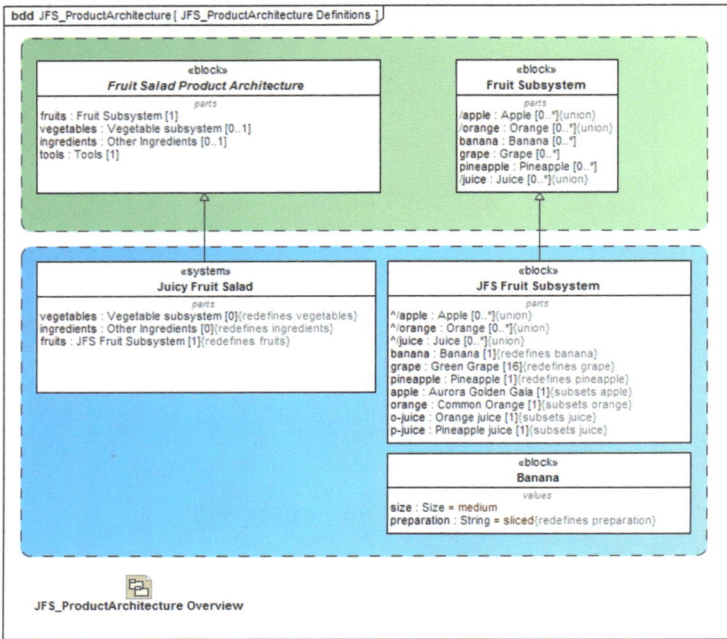

Figure 3.36: Definition of the Juicy Fruit Salad product architecture

The *Banana* is not part of the core model and added in the variant configuration. It is an example where the modeler has decided not to add a block to the core. The element only exists in the variant configuration. Another source of the *Banana* could be a variant or a fruit model library. However that is not modeled here.

Interesting to note that the properties are just properties without any association relationships. In block modeling with SysML you can often skip the modeling of association and model the part properties directly. That reduces the modeling effort of creating and maintaining the models. Unfortunately, that is formally forbidden in the current SysML version 1.4. Each property that is typed by a block must be defined by an association. However most SysML modeling tools ignore that constraint and there is an open issue submitted to the OMG to remove the constraint in a future version of SysML (see We16)

Figure 3.37 shows instance specifications of a juicy fruit salad. They are automatically created by the modeling tool and, for example, could be used to verify whether the structural definitions are correct.

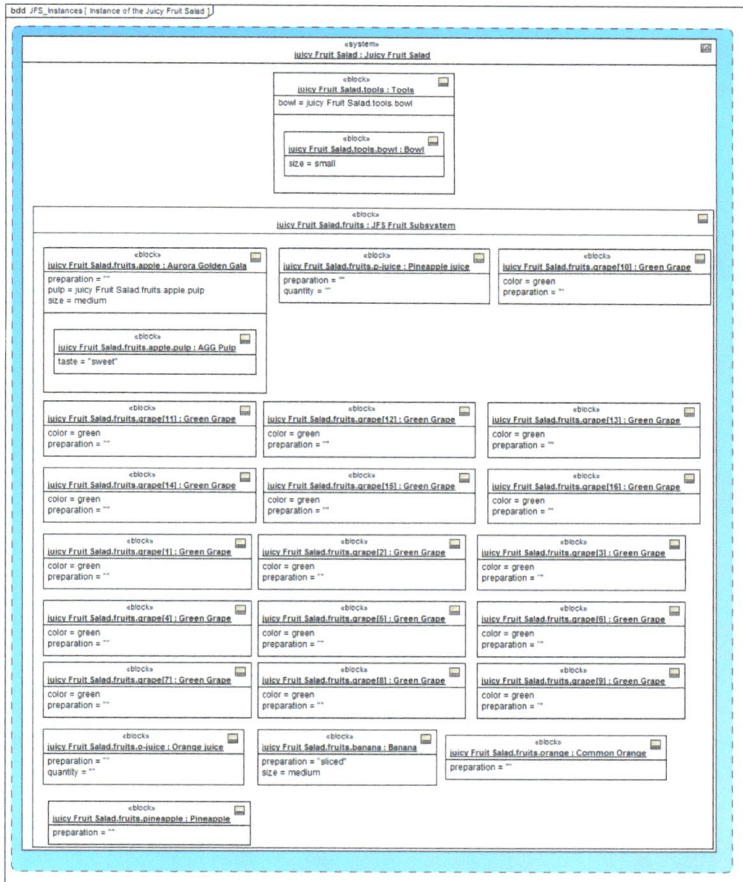

Figure 3.37: Instance Specifications of the Juicy Fruit Salad

Figure 3.38 depicts another variant configuration respectively fruit salad recipe. The modeling is similar to the juicy fruit salad and will not be further discussed here in detail. You can find the complete variant configuration in the accompanying model at www.model-based-systems-engineering.com/vamos.

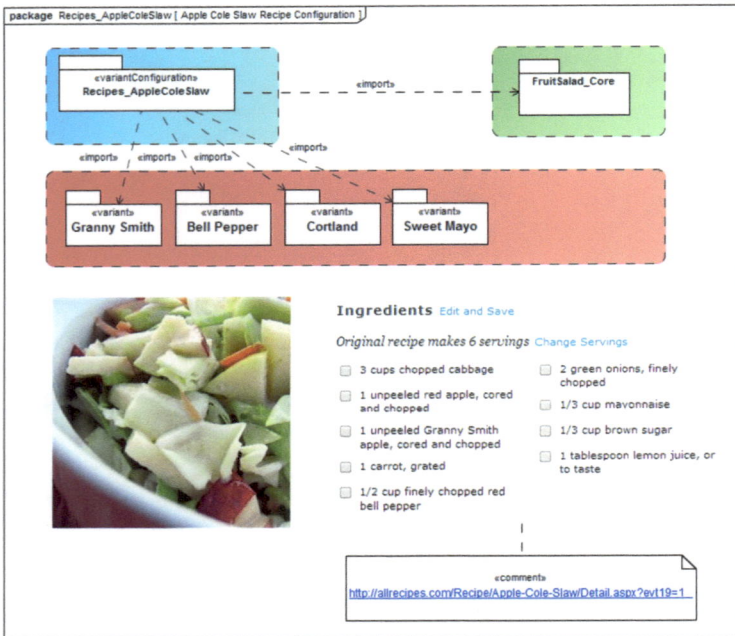

Figure 3.38: Recipe (variant configuration) Apple Cole Slaw

Now we have modeled all elements of the core, variants, and a concrete variant configuration. Figure 3.39 provides a brief overview of the relationships of the VAMOS approach.

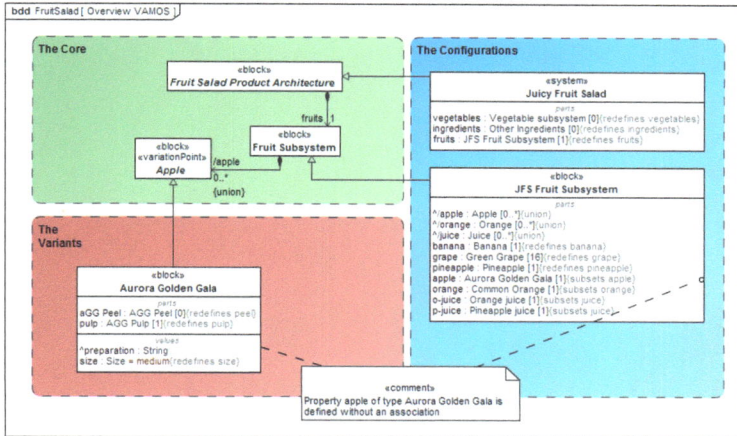

Figure 3.39: Overview of VAMOS relationships

Although all these stereotypes are simple and powerful it is a challenge to handle the complexity of the model. Even a system model without variants could already be a challenge. With variants it is a multidimensional configuration space. Special views, reports and model transformations are necessary to manage the complexity.

4. Other Variant Modeling Concepts

In this chapter I look out of the box and describe some other variant modeling concepts. By now there are only three different concepts and they are only briefly described. I will provide a more detailed and broader analysis of other variant modeling concepts in future editions of this book.

4.1 Feature Oriented Domain Analysis (FODA)

A common and old concept to model variability is the *Feature Oriented Domain Analysis (FODA)* by Kang et al. (Ky90). The variability is modeled from the perspective of the stakeholders. It shows the features of the system, their variability and constraints between the variants.

Figure 4.1 shows an extract of the FODA feature tree of the Fruit Salad. It depicts the three features *Pepper*, *Apple*, and *Dressing*. The lines with a small circle at one end depict optional features. Without the circle it is a mandatory feature. The arc between two lines depicts that the related features are exclusive and could not be part of the same product at the same time.

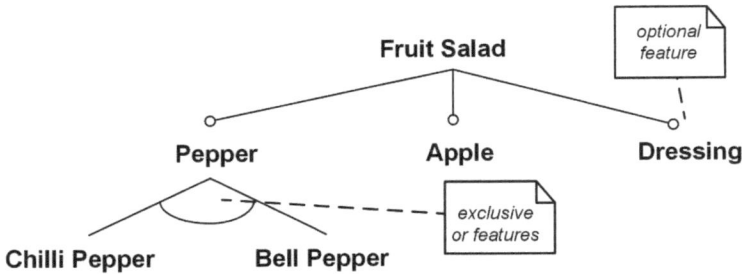

Figure 4.1: Example FODA feature tree of the Fruit Salad

The FODA feature tree is the mother of the feature trees. The concept is used in many variant descriptions and modeling approaches.

Later Kang extended FODA to the *Feature-Oriented Reuse Method (FORM)* that in addition to the requirements and architecture perspective incorporates the marketing perspective (Ky02).

4.2 Common Variability Language (CVL)

The *Common Variability Language (CVL)* is a concept and language that was planned to be an adopted standard of the OMG (CVL16). Unfortunately the adoption process has stopped and it seems that CVL will not become a standard anymore.

CVL is a modeling language to specify the variability aspects of any model that is defined based on the Meta Object Facility (MOF) (MOF15).

Figure 4.2 sketches the notion of CVL. The base model describes the system of interest. For example, a system model in SysML. The variability model describes the variability aspects of the base model in CVL. And the resolution model describes how to resolve the variability in CVL. A model-to-model transformation creates the resolved model in the same language as the base model, for example SysML.

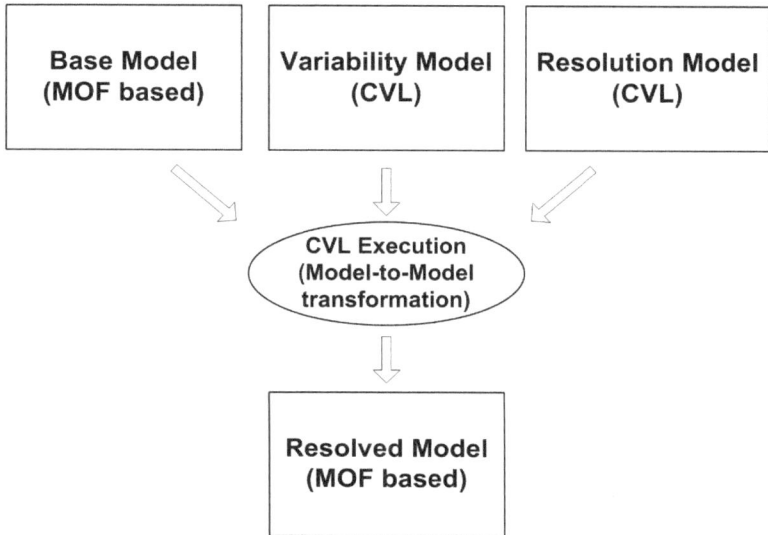

Figure 4.2: Concept of the Common Variability Language

CVL provides two levels of modeling. On the user-centric layer the modeler can do high level variability modeling based on features. It is similar to the FODA feature tree.

On the product-realization layer the modeler specifies the details of the variability.

4.3 Orthogonal Variability Model (OVM)

The *Orthogonal Variability Model (OVM)* is described by Pohl et. al in Poh05. An orthogonal variability model captures the variability of software artifacts in a separate model. OVM has its roots in software engineering, but could also be applied to systems engineering models.

Also OVM provides a feature tree as shown in figure 4.3. The node with the VP symbol represents a variation point. The nodes with the V symbol represent variants. The dashed lines depict optional variants. Mandatory variants are depicted by solid lines.

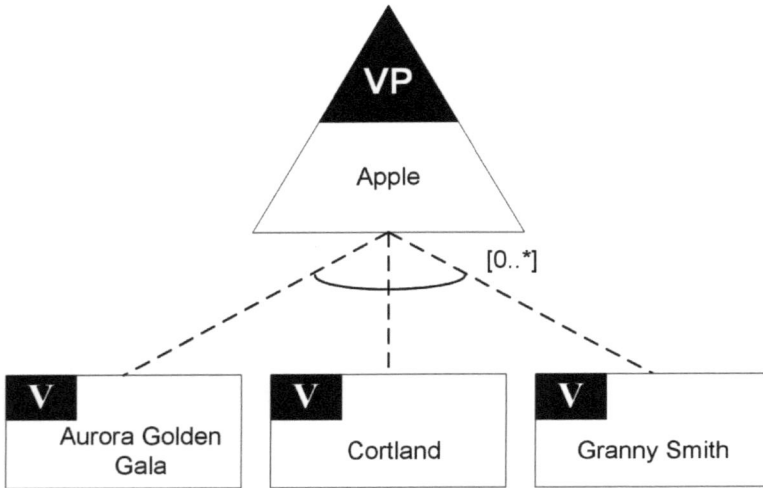

Figure 4.3: Apple variation point and variants with OVM

5. Variant Stereotypes for SysML

This chapter presents the definitions of the VAMOS stereotypes. They are part of the SYSMOD profile (We15). You can use them independently of the other SYSMOD stereotypes, although I recommend the SYSMOD toolbox to create your system model.

Figure 5.1 depicts all VAMOS stereotypes in a single diagram. In the following each stereotype is described in subchapters in alphabetical order. See also chapter 2 for the semantic of the variant concepts implemented by the VAMOS stereotypes.

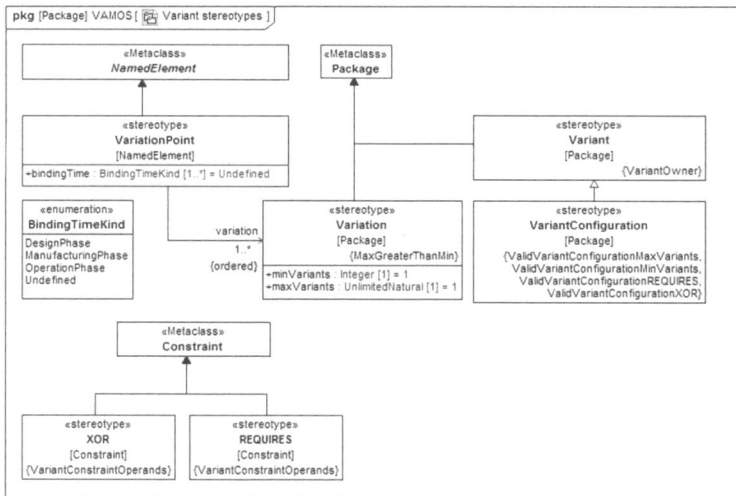

Figure 5.1: VAMOS stereotypes of the SYSMOD profile

5.1 BindingTimeKind

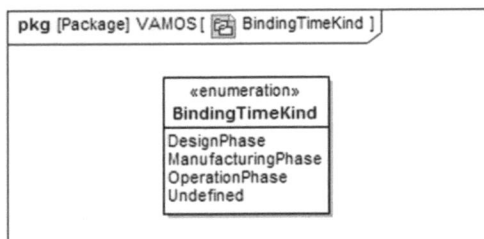

Figure 5.2: Enumeration BindingTimeKind

BindingTimeKind is an enumeration type with four enumeration literals to specify the time when a variant is bound to a variation point (figure 5.2). The time is the latest point when a decision for the variant must be made.

DesignPhase	The variant is bound during the design phase to the variation point. It is possible to adapt the design for the selected variant, for example to optimize the system for strong performance requirements.
ManufacturingPhase	The variant is bound during the manufacturing of the system.
OperationPhase	The variant is bound during the operation of the system, for example by a software update or replacement of a system component.
Undefined	The binding time is not defined.

The enumeration type is used by a property of the VariationPoint stereotype.

5.2 REQUIRES constraint

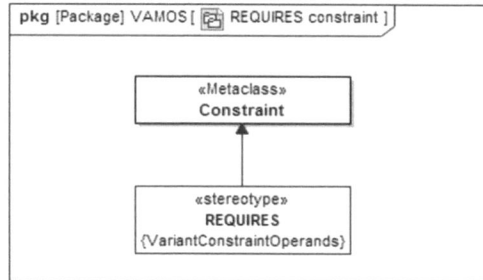

Figure 5.3: Stereotype REQUIRES

The REQUIRES constraint is a predefined binary constraint between two variants. It specifies that the second variant must also be selected for a valid variant configuration if the first variant is selected. The validation of the REQUIRES constraint is defined in a constraint applied to the VariantConfiguration stereotype (section 5.4).

The operands of the *REQUIRES* constraint are stored in the property *constrainedElement* owned by the SysML *Constraint* element. It is an ordered set and thus clearly defines that the first variant requires the second variant in the ordered set for a valid variant configuration.

Figure 5.3 depicts the stereotype definition of the *REQUIRES* variant constraint. The REQUIRES constraint itself has a constraint that assures that it has exactly two operands (*constrainedElement*) of type Variant. The constraint *VariantConstraintOperands* is defined with OCL[1] as follows:

---

[1]Object Constraint Language (OCL14).

```
context REQUIRES inv VariantConstraintOperands:
self.constrainedElement->size()=2 and self.constrainedElement->asSet\
()->forAll(oclIsTypeOf(SYSMOD::VAMOS::Variant))
```

5.3 Variant

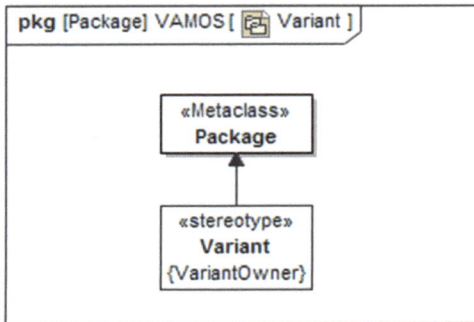

Figure 5.4: Stereotype Variant

The *Variant* stereotype is applied to a package that contains all specific elements of the variant. The package structure within the variant package is the same as for the system or a system component. In particular, the variant could also have variant configurations and variations with variants.

A variant must be owned by a *Variation* (section 5.5). That rule is assured by the constraint *VariantOwner* that is applied to the variant stereotype. The constraint is defined with OCL as follows:

```
context Variant inv VariantOwner:
self.owningPackage.oclIsTypeOf(SYSMOD::VAMOS::Variation)
```

5.4 VariantConfiguration

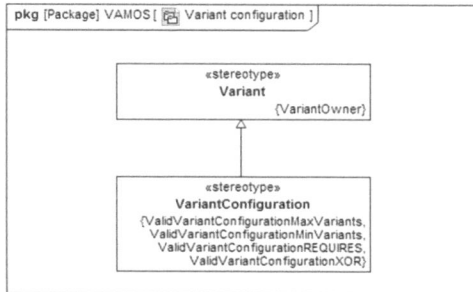

Figure 5.5: Stereotype Variant configuration

A *Variant configuration* is a special kind of a *Variant*. It defines a system or system component by a set of *Variants*, the core, glue logic elements for the connection of the components, and - if applicable - own system elements.

A variant configuration imports the core and selected Variants by using the SysML *PackageImport* relationship.

The constraints *ValidVariantConfigurationMinMaxVariants, ValidVariantConfigurationREQUIRES*, and *ValidVariantConfigurationXOR* assures that the selected set of imported Variants is valid, i.e. the *minVariants/maxVariants* definition and XOR and REQUIRES constraints are fulfilled.

The constraints are defined with OCL as follows:

The constraints *ValidVariantConfigurationMinMaxVariants* and *ValidVariantConfigurationMaxVariants* check for each variant configuration that the value of *minVariants* of all included variations is less or equal than the number of imported variants of the same variation and appropriately for *maxVariants*.

```
context VariantConfiguration inv ValidVariantConfigurationMinVariant\
s:
self.packageImport.target->select(oclIsTypeOf(SYSMOD::VAMOS::Variant\
)).owner
->forAll( v |
v.oclAsType(SYSMOD::VAMOS::Variation).minVariants <=
v.oclAsType(Package).ownedElement->intersection(self.packageImport.t\
arget->select(oclIsTypeOf(SYSMOD::VAMOS::Variant)))->size()
)
```

```
context VariantConfiguration inv ValidVariantConfigurationMaxVariant\
s:
self.packageImport.target->select(oclIsTypeOf(SYSMOD::VAMOS::Variant\
)).owner
->forAll( v |
v.oclAsType(Package).ownedElement->intersection(self.packageImport.t\
arget->select(oclIsTypeOf(SYSMOD::VAMOS::Variant)))->size() <=
v.oclAsType(SYSMOD::VAMOS::Variation).maxVariants
)
```

The constraint *ValidVariantConfigurationREQUIRES* checks for each variant configuration that all variants that are required by the REQUIRES constraint of imported variants are a member of the selected set of variants.

```
context VariantConfiguration inv ValidVariantConfigurationREQUIRES:
let cset:Set(Constraint)=self.importedMember->select(oclIsKindOf(SYS\
MOD::VAMOS::REQUIRES))->collect(oclAsType(Constraint))->asSet()  in
cset->forAll(self.packageImport.target->asSet()->includesAll(constra\
inedElement->excluding(owner)->asSet()))
```

The constraint *ValidVariantConfigurationXOR* checks for each variant configuration that all variants that are connected by imported variants with a XOR constraint are not a member of the selected set of variants.

```
context VariantConfiguration inv ValidVariantConfigurationXOR:
let cset:Set(Constraint)=self.importedMember->select(oclIsKindOf(SYS\
MOD::VAMOS::XOR))->collect(oclAsType(Constraint))->asSet()  in
cset->forAll(constrainedElement->excluding(owner)->asSet()->intersec\
tion(self.packageImport.target)->size()=0)
```

5.5 Variation

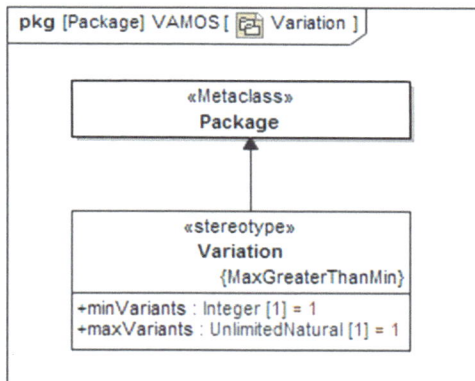

Figure 5.6: Stereotype Variation

The *Variation* stereotype is applied to a package that contains *Variants*. It represents the discriminator of the included *Variants*. For example, the variation *GearKind* is the discriminator of the *Variants Manual* and *Automatic* in a car system model.

The properties *minVariants* and *maxVariants* specify the minimum and maximum number of Variants that could be selected from this variation for a single valid Variant Configuration. The default value for both properties is 1, i.e. exactly one Variant must be selected. For example, a car must have exactly one gear kind - automatic or manual.

The constraint *MaxGreaterThanMin* applied to the variation stereo-type assures that the *maxVariants* value is always greater or equal than the *minVariants* value. The constraint is defined with OCL as follows:

```
context Variation inv MaxGreaterThanMin:
self.minVariants <= self.maxVariants
```

5.6 VariationPoint

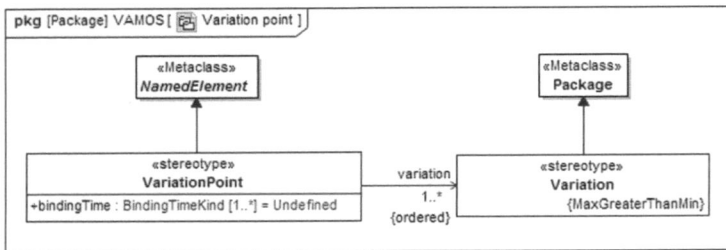

Figure 5.7: Stereotype Variation point

A *VariationPoint* marks a model element in the core that exists in different Variants. It could be any model element that has a name. The stereotype extends the abstract metaclass *NamedElement*.

The property *bindingTime* is described in section 5.1.

The property *variants* specifies a set of at least one *Variation* that contains the variant elements that could docks at the variation point. If the variation point is a classifier like a *Block* the relationship between the variant element and the variation point is typically the *Generalization* relationship.

5.7 XOR constraint

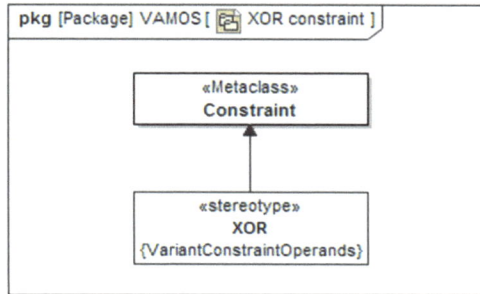

Figure 5.8: Stereotype XOR

The XOR constraint is a predefined binary constraint between two variants. It specifies that the two related variants could not be part of the same valid variant configuration. The validation of the XOR constraint is defined in another constraint applied to the VariantConfiguration stereotype.

The XOR constraint itself has a constraint that assures that it has exactly two operands (*constrainedElement*) of type Variant. The constraint *VariantConstraintOperands* is defined with OCL as follows:

```
context XOR inv VariantConstraintOperands:
self.constrainedElement->size()=2 and self.constrainedElement->asSet\
()->forAll(oclIsTypeOf(SYSMOD::VAMOS::Variant))
```

Appendix A - Example Forest Fire Detection System

The Forest Fire Detection System (FFDS) observes a defined area for forest fires. The core use case of the system is to report a fire within a defined time frame from the burst of the fire. The example is taken from the book SYSMOD - The Systems Modeling Toolbox - Pragmatic MBSE with SysML (We15).

Section A.1 presents the core of the FFDS and gives a good overview of the system. The following sections describe the variant aspects of the FFDS.

A.1 The Core

There are many implementations possible to provide a forest fire detection service: satellites (e.g. (JPL16)), forest animals equipped with sensors (Sa07), drones, watch towers, planes, and more. To narrow the solution space we define a base architecture that presets some architecture and technical decisions. Figure A.1 depicts a sketch of the base architecture on a beer mat. Figure A.2 shows a more formal version in a SysML internal block diagram.

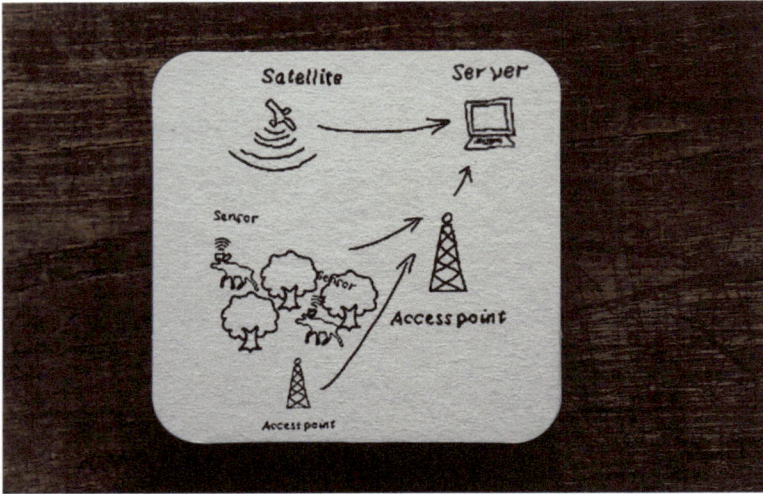

Figure A.1: Sketch of the Base Architecture of the FFDS on a beer mat

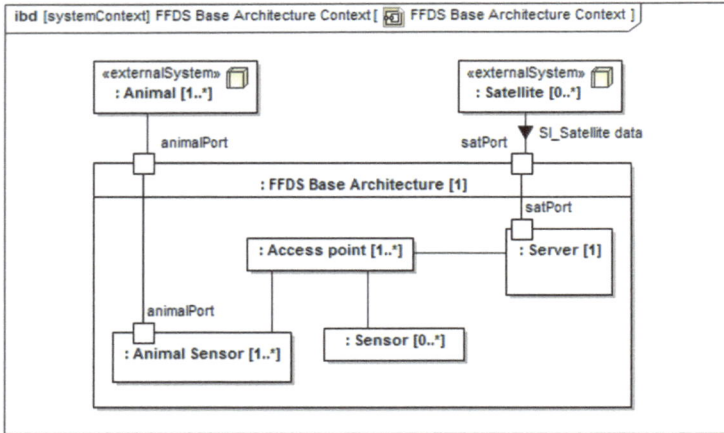

Figure A.2: Base Architecture of the FFDS

The base architecture presets the option for a satellite observation. The satellites are external to the system. The animal sensors are the special feature of the system and a mandatory part. The sensors are connected with access points to communicate with a central server.

Figure A.3 shows the system context of the FFDS. It is a specialization of the base architecture context and is derived from the user requirements.

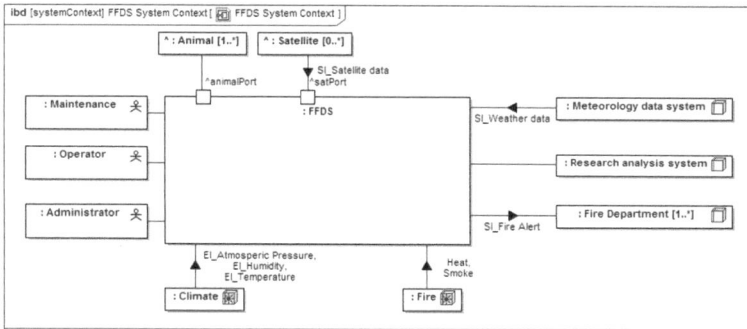

Figure A.3: FFDS System Context

Figure A.4 depicts the main use cases of the FFDS operator.

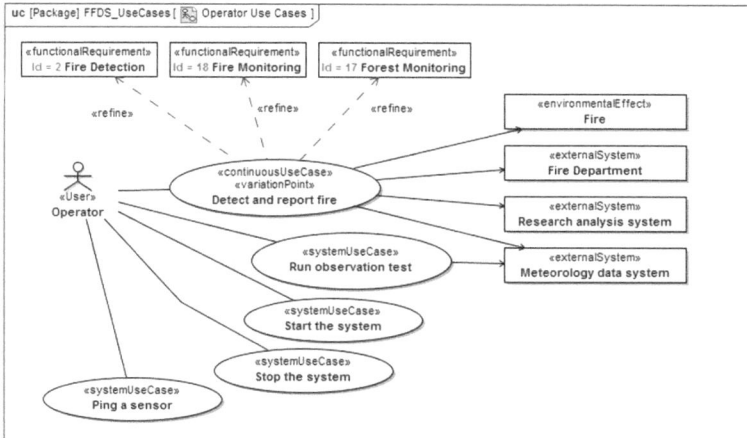

Figure A.4: FFDS Operator Use Cases

The logical architecture of the FFDS is shown in figure A.5. The
main elements are the logical unit of all sensors installed in the
forest and the central control system. There are two central control
systems (multiplicity 2): the primary one and a backup system.

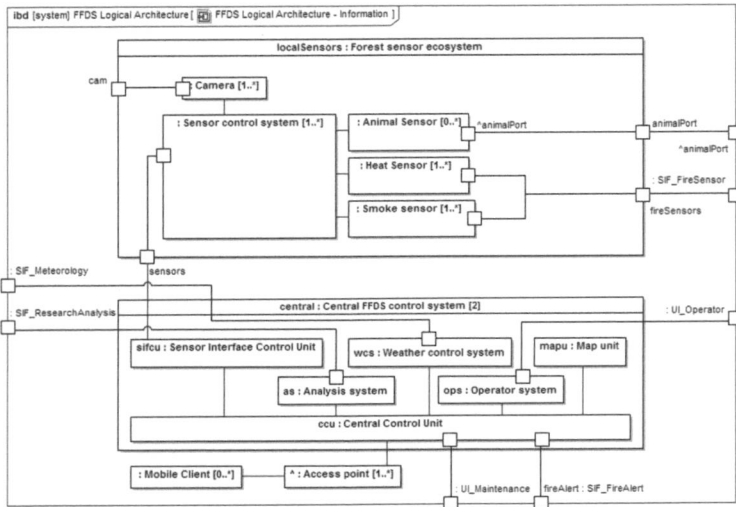

Figure A.5: Logical Architecture of the FFDS

A.2 The Variants

The feature tree in figure A.6 gives a good overview about the
variants of the FFDS. There is only one variation with variants for
additional forest fire detection subsystems. The satellite subsystem
is initially specified in the base architecture. The core has only
placeholders for the satellite respectively the appropriate variation
points. The detailed specification is contained in the variant.

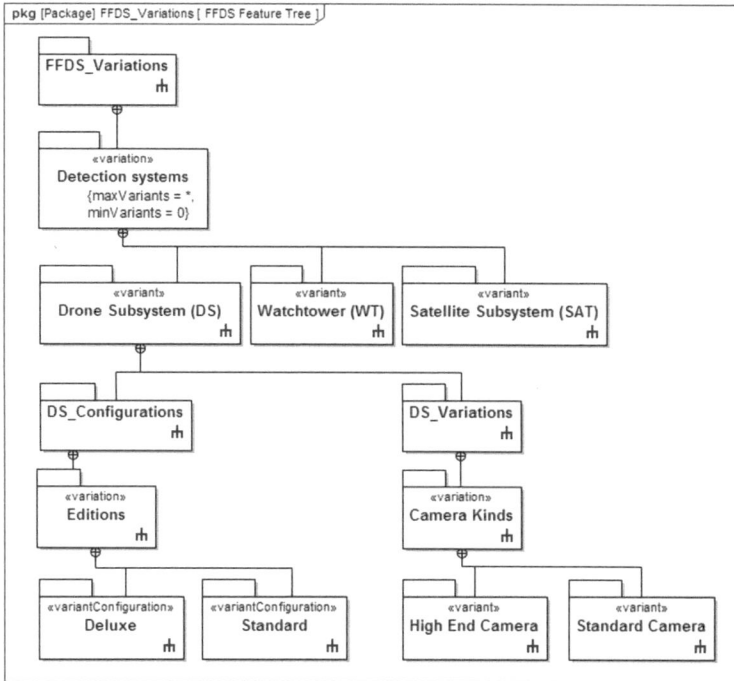

Figure A.6: Feature Tree of the FFDS

Figure A.7 depicts the definition of the central control system for the satellite subsystem variant. It is a specialization of the core central control system with the addition of a satellite control system.

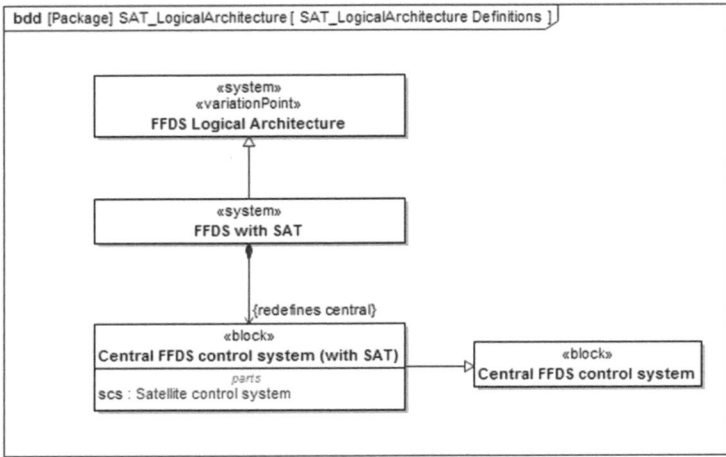

Figure A.7: Control System with satellite support

The drone subsystem is another kind of detection subsystem. It is available in two different configurations: the deluxe edition comes with drones with high end cameras and the standard edition comes with drones with standard cameras.

The watchtower is a simple observation platform for a human observer with a client application to communicate with the server of the FFDS. The watchtower variant introduces a new actor *Forest Observer*. Figure A.8 shows the specialized system context.

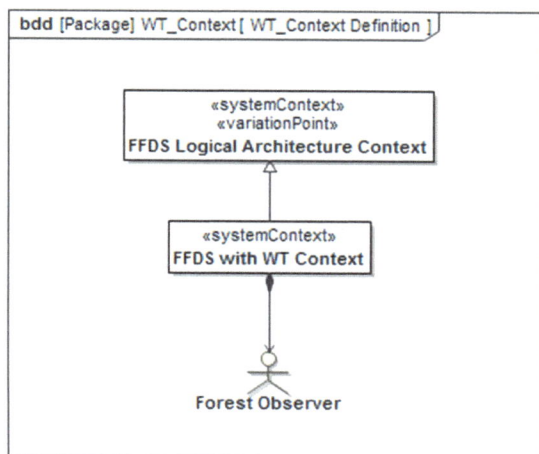

Figure A.8: System context of the Watchtower variant

A.3 The Variant Configurations

The FFDS is available in different configurations

- A version without satellites called *Standard*.
- A version without satellites, but with watchtowers called *Standard Observer*.
- A version with satellite support called *Premium*.
- The *Standard* as well as the *Premium* version is available in combination with a drone surveillance system.

Figure A.9 depicts the different variant configurations in a variant configuration tree.

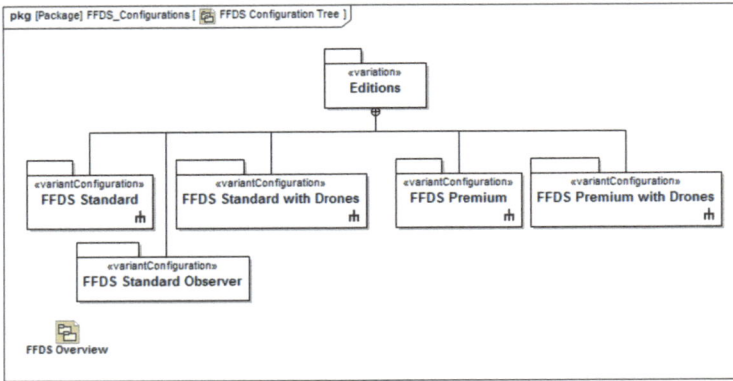

Figure A.9: FFDS Variant Configurations Tree

Next I put a focus on the variant configuration *FFDS Standard,* i.e. a system without satellite observation, without watchtowers, and without a drone surveillance system. The satellite observation option is part of the base architecture and the core. Therefore the variant configuration must explicitly remove the satellite option. Typically the satellite should not be part of the core, because it is not part of all variant configurations. However here it is a decision of the project, that the satellite is part of the base architecture in the core.

Figure A.10 shows the system context of the *FFDS Standard* that removes the satellite actor and port from the configuration by setting the multiplicity to 0.

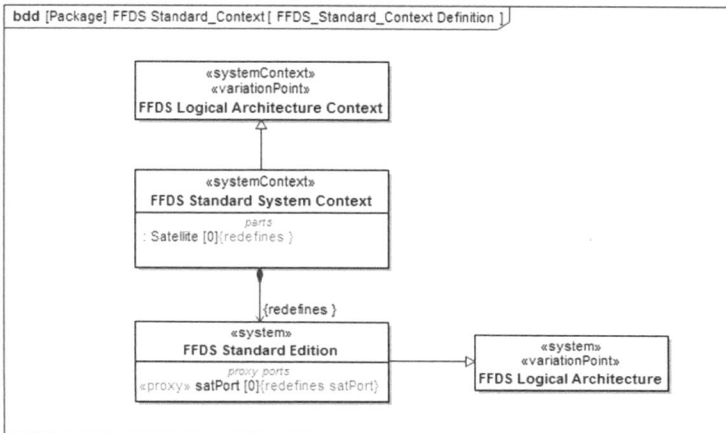

Figure A.10: FFDS Standard Edition System Context

The variant configuration *Standard Observer* is similar to the *Standard* configuration. Therefore it reuses the context and system element as depicted in figure A.11.

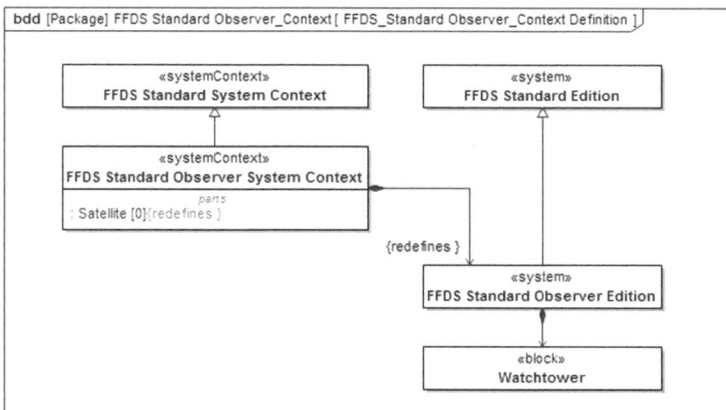

Figure A.11: FFDS Standard Observer Edition System Context

Figure A.12 shows the definition of the system context element of the premium edition with satellite and drone support (drone deluxe edition). The variant configuration is similar to the *FFDS Premium* configuration. Therefore that configuration is reused as depicted in figure A.12.

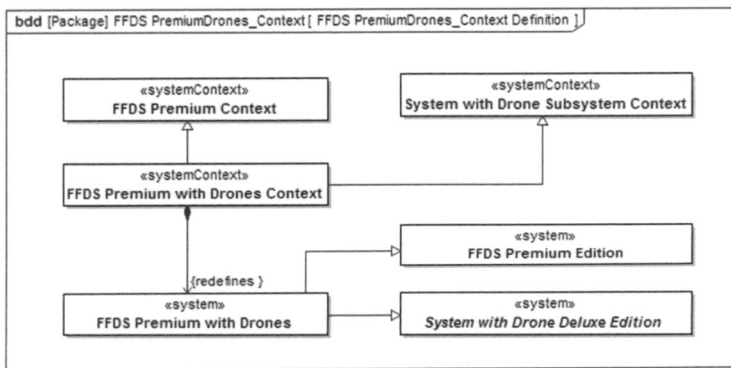

Figure A.12: FFDS Premium with Drones Edition System Context

Appendix B - Example Virtual Museum Tour System

The example of the Virtual Museum Tour (VMT) is taken from my book *Model Based System Architecture* (WeLa15) that I have written together with Jesko G. Lamm, Stephan Roth and Markus Walker. The VMT system provides virtual visits of a museum especially outside the opening hours. The customer watches the museum through the eyes of a robot. The system provides fix tours where the customer just watches the video stream from the robot whiles it moves automatically through the museum and listens to the explanations from a tour guide. The system also provides custom virtual museum tours where the customer can control the robot and steer it through the museum.

Section B.1 gives a brief overview of the core including a description of the purpose of the VMT. More details of the example can be found in WeLa15. The following sections describe the variant aspects of the VMT.

Note that the example presented here slightly differs from the example in WeLa15. The books have both their own lifecycle and herewith their own example model.

B.1 The Core

This section presents the main artifacts of the core of the VMT to get an overview of the system.

Figure B.1 depicts the system context of the VMT including the top level of the logical architecture embedded in the system model element.

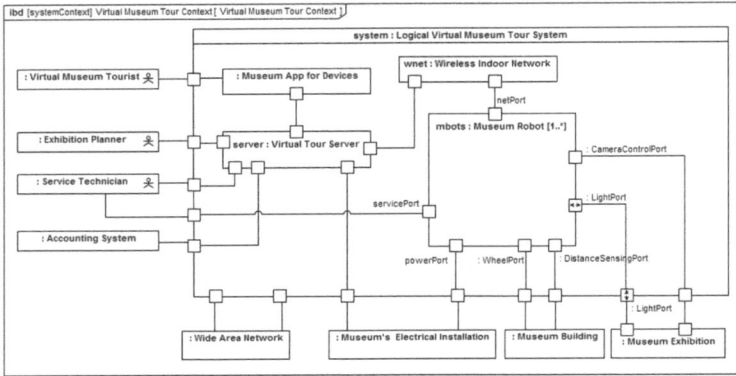

Figure B.1: VMT System Context

The main actor of interest is the customer of the VMT: the Virtual Museum Tourist. The actor is depicted on the upper left side of the system context diagram. The main components of the logical architecture are a virtual tour server, the museum robots and an app for the user devices, for example a smart phone or laptop.

The app is the user interface of the VMT for the virtual museum tourist. The app shows the museum visit, provides control elements for the robots, and some general functions like settings, user profiles, and the user payment data.

The museum robots are the eyes of the virtual museum tourists. They move through the museum and stream videos. The robots can be steered by the virtual museum tourist by using the app or they move automatically controlled by the virtual tour server.

The server is also responsible for managing the museum tours, and to manage the customer profiles and payments. It is the central control unit of the VMT.

Figure B.2 depicts some of the use cases of the virtual museum tourist and gives a good overview of the services provided by the VMT.

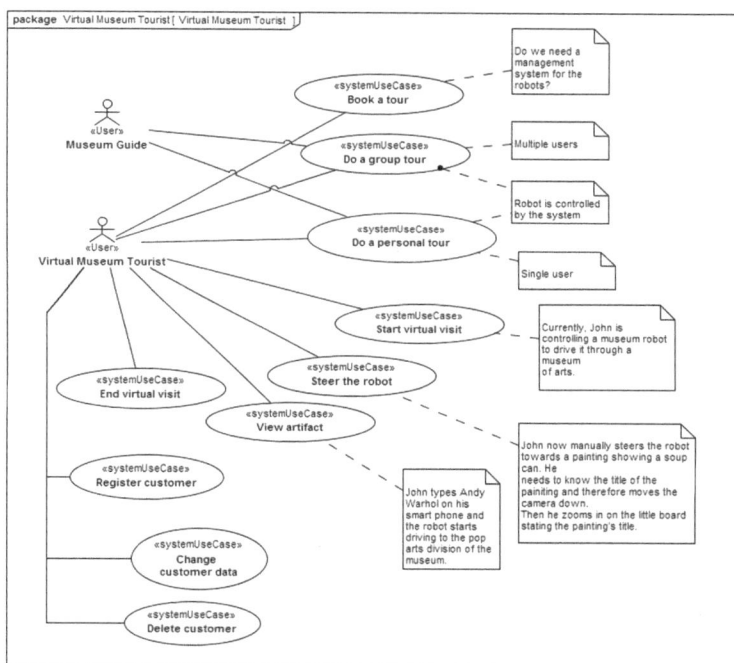

Figure B.2: VMT Use Cases

Figure B.3 shows the top level package structure of the VMT model. The structure follows the guideline for system model package structures of the SYSMOD toolbox (We15).

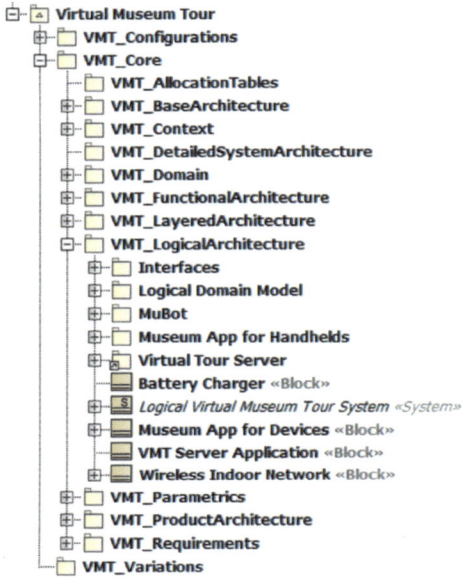

Figure B.3: VMT Package Structure

Figure B.4 depicts the package structure of the logical architecture. Some blocks of the architecture are organized in packages with a substructure similar to the system package structure, for example the package for the museum robots (*MuBot*). This recursive pattern of the package structure makes the model scalable for very large system model and enables the reuse of a single system assembly.

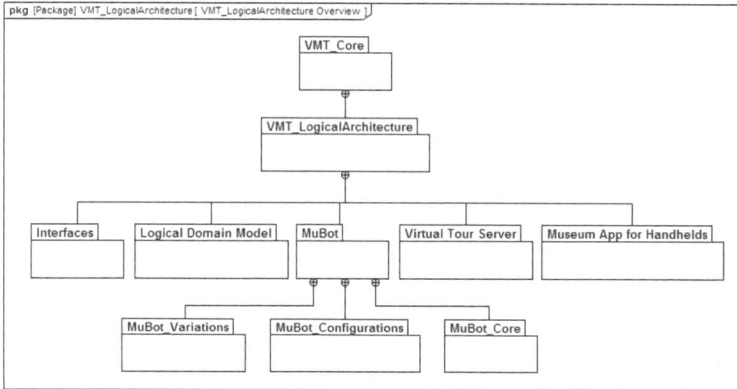

Figure B.4: Package structure of the logical architecture

B.2 The Variants

The VMT system itself has no variants. You may have noticed in figure B.3 that the package *VMT_Variations* has no leading plus sign, i.e. it is empty. Of course the system includes variants; otherwise I would not have selected it as an example for variant modeling. The variants are embedded in the logical architecture of the system.

Figure B.5 depicts the substructure of the *MuBot* package (museum robots). You can see a variations package with variations *MuBot_-AntiCollision, MuBot_Chassis*, and *MuBot_VisualSystem*.

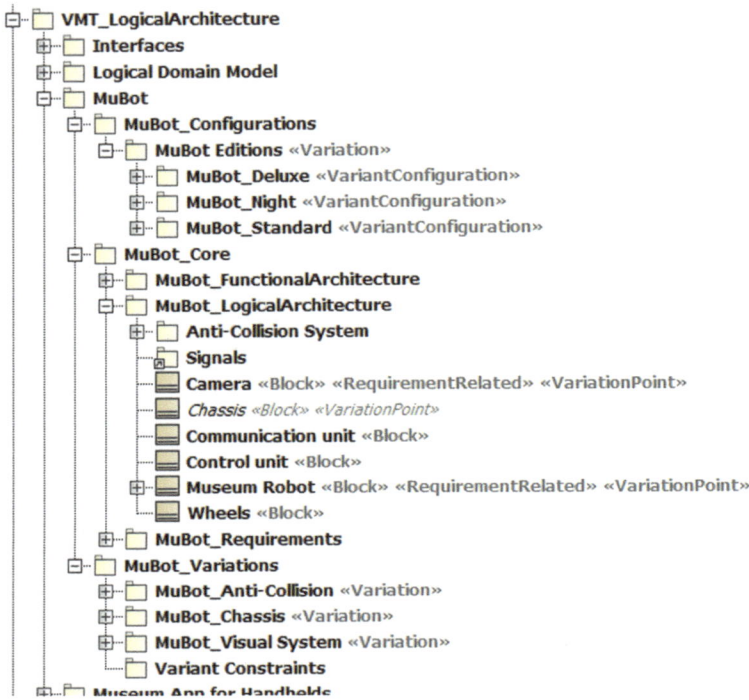

Figure B.5: VMT MuBot Package Structure

Figure B.6 depicts the feature tree of the *MuBot*. A museum robot must have an anti-collision system (ACS) to avoid collisions with things, walls, and persons in the museum. Therefore the value of the variation property *minVariants* is 1. There are two variants available: a camera-based ACS and an ultrasound-based ACS.

The chassis variation includes two variants. For museum robots that operate outside the opening hours of the museum it is sufficient to be equipped with the standard chassis. The deluxe chassis has a much better look and could be adapted to the kind and brand of the museum. It is well-suited for museum robots that operate during the opening hours of the museum and are seen by the real physical museum visitors.

The variation of the visual system includes cameras with and without a light system. The camera with lights again includes a variation for the kind of the light system. The special light variant is a light system to protect sensitive museum artifacts. This is an example of nested variants (see also figure 2.2).

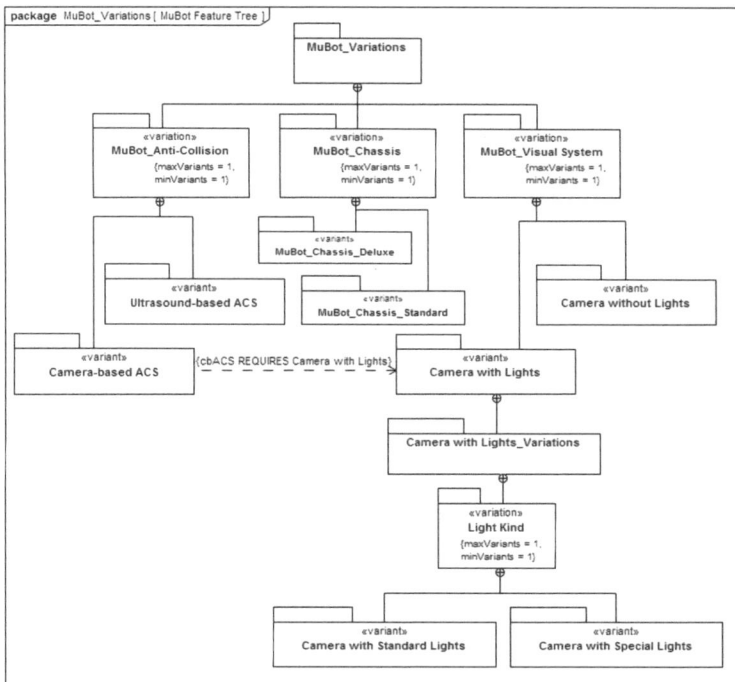

Figure B.6: MuBot Feature Tree

Figure B.7 depicts the definition of *MuBot* blocks. The *Camera, Chassis* and the *Anti-Collision System* are variation points.

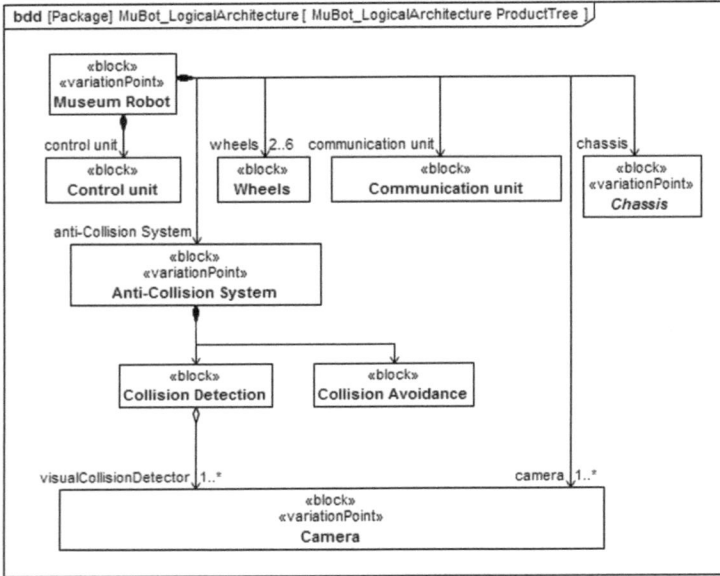

Figure B.7: Definition of the museum robot

Figure B.8 shows how variant elements specialize the variation points of the museum robot.

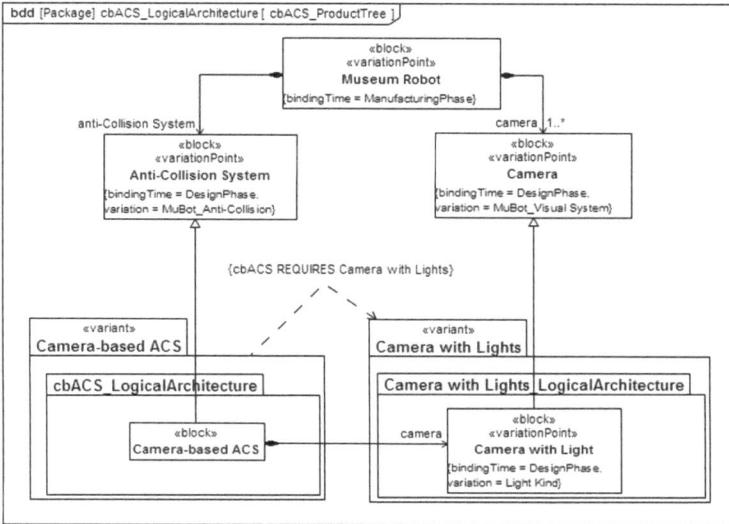

Figure B.8: Logical Architecture of the Anti-collision System

The definition of the variant *Camera with special lights* is shown in figure B.9. The special lights are lights that do not harm sensitive museum artifacts.

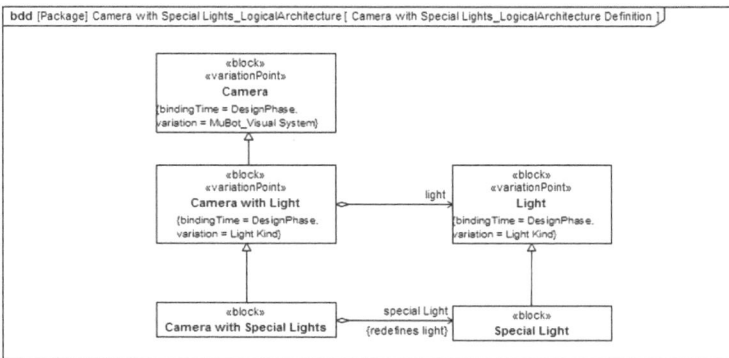

Figure B.9: MuBot camera with special lights

B.3 The Variant Configurations

Figure B.10 depicts the configuration tree of the MuBot. There are three variant configurations: *MuBot_Standard, MuBot_Night,* and *MuBot_Deluxe.* In the following I show some snippets from the *MuBot_Deluxe* configuration.

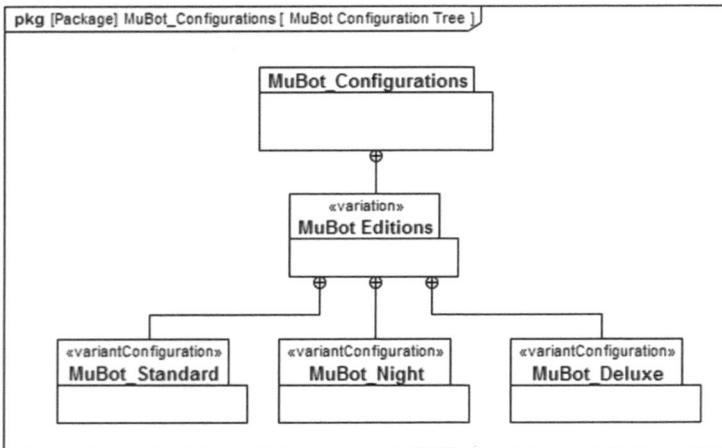

Figure B.10: MuBot Configuration Tree

Figure B.11 depicts the set of variants selected for the deluxe museum robot. Of course it has selected the deluxe chassis. Additionally the ultrasound-based anti-collision system and the camera system with the special lights are part of the configuration.

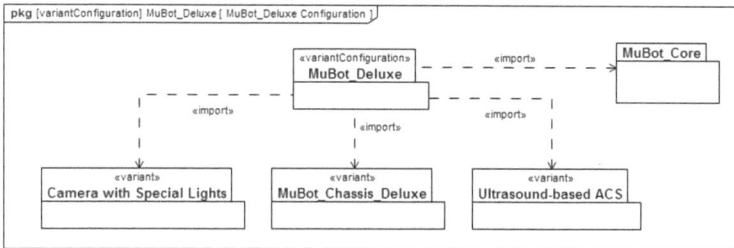

Figure B.11: MuBot Deluxe Configuration

The definition of the *MuBot Deluxe* block is shown in figure B.12. The three depicted part properties redefine the appropriate properties of the variation point *MuBot*.

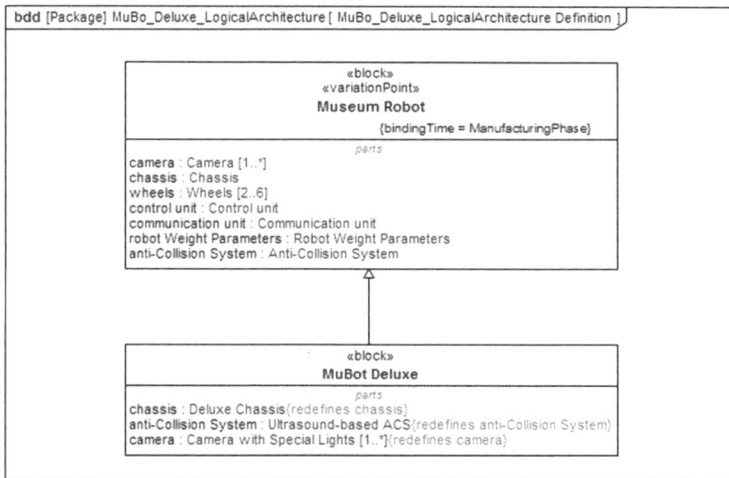

Figure B.12: MuBot Deluxe

Figure B.13 depicts the configuration tree of the whole virtual museum tour system.

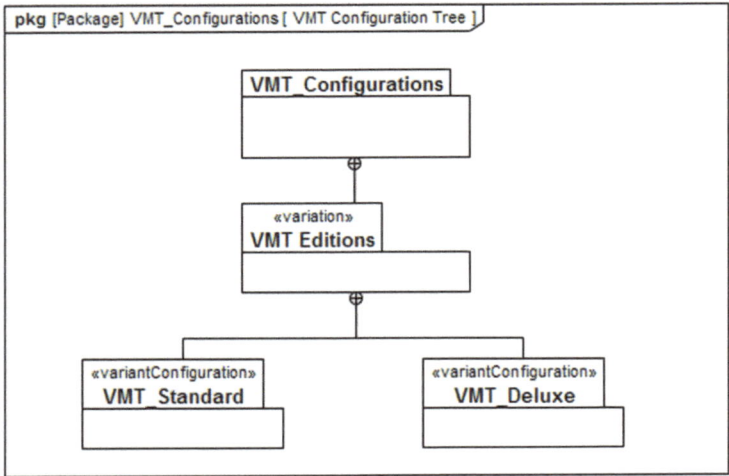

Figure B.13: VMT Configuration Tree

The selection of the *VMT_Deluxe* configuration is depicted in figure B.14.

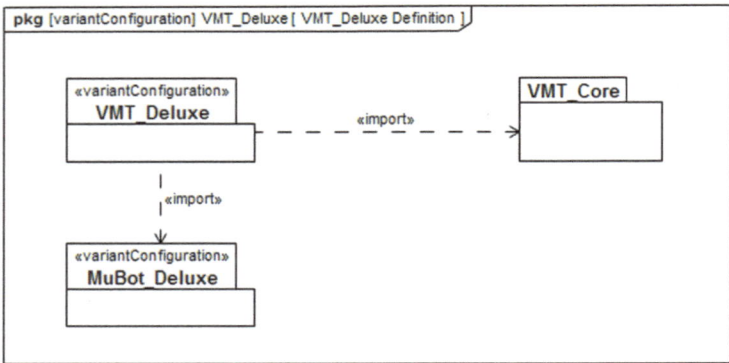

Figure B.14: Variant selections of configuration *VMT_Deluxe*

Figure B.15 shows the definition of the *VMT Deluxe* system. It redefines the *mbots* property of the core system element to change the type to the deluxe museum robot.

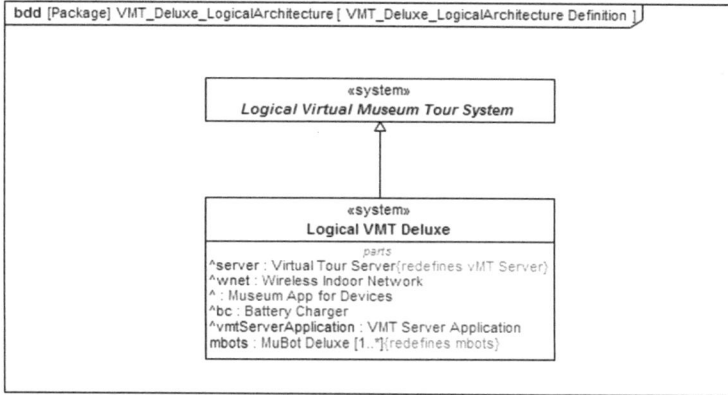

Figure B.15: Definition of the *VMT_Deluxe* system

Bibliography

(Al16) allrecipes.com, accessed January 2016.

(CVL16) http://www.omgwiki.org/variability/doku.php. accessed March 2016.

(JPL16) http://wildfire.jpl.nasa.gov/. accessed January 2016.

(Ky90) Kyo C. Kang, Sholom G. Cohen, James A. Hess, William E. Novak, and A. Spencer Peterson. *Feature-oriented domain analysis (foda) feasibility study.* Technical Report CMU/SEI-90-TR-21, Software Engineering Institute, 1990.

(Ky02) Kyo C. Kang, Jaejoon Lee, and Patrick Donohoe. *Feature-oriented product line engineering.* IEEE Software, 19:58-65, 2002.

(MOF15) Object Management Group. Meta Object Facility (MOF) Version 2.5. 2015.

(OCL15) Object Management Group. Object Constraint Language (OCL) Version 2.4. 2014.

(Poh05) Klaus Pohl, Günter Böckle, and Frank J. Linden. *Software Product Line Engineering: Foundations, Principles and Techniques.* Springer, 1 edition, 2005.

(Sa07) Sahin YG. Animals as Mobile Biological Sensors for Forest Fire Detection. Sensors (Basel, Switzerland). 2007;7(12):3084-3099.

(Sy15) Object Management Group. *OMG Systems Modeling Language (OMG SysML) - Version 1.4.* formal/2015-06-03.

(We08) Tim Weilkiens. *Systems Engineering with SysML/UML.* Morgan Kaufmann, 2008.

(We14) Tim Weilkiens. *Systems Engineering mit SysML/UML.* 3. Auflage, dpunkt-Verlag, 2014.

(WeLa15) Tim Weilkiens, Jesko G. Lamm, Stephan Roth, Markus Walker. *Model-Based System Architecture*. Wiley, 2015.

(We15) Tim Weilkiens. *SYSMOD - The Systems Modeling Toolbox*. Leanpub, 2015.

(We16) Tim Weilkiens. *The Myth of the Association*. Blogpost www.model-based-systems-engineering.com, 2015.

Index